GARDEN REVOLUTION

GARDEN REVOLUTION

How our landscapes can be a source of environmental change

LARRY WEANER
THOMAS CHRISTOPHER

TIMBER PRESS
PORTLAND, OREGON

Page 2: This pond and waterfall were designed to appear as though they were naturally occurring. Page 6: A wasp visits Indian plantain.

Published in 2016 by Timber Press, Inc.

The Haseltine Building
133 S.W. Second Avenue, Suite 450
Portland, Oregon 97204-3527
timberpress.com

Printed in China
Text design by Debbie Berne
Jacket design by Anna Eshelman

A catalog record for this book is also available from the British Library.

Library of Congress Cataloging-in-Publication Data

Names: Weaner, Larry, author. | Christopher, Thomas, author.
Title: Garden revolution: how our landscapes can be a source of environmental
    change / Larry Weaner and Thomas Christopher.
Description: Portland, Oregon: Timber Press, 2016. | Includes bibliographical
    references and index.
Identifiers: LCCN 2015036650 | ISBN 9781604696165
Subjects: LCSH: Landscape ecology. | Gardening.
Classification: LCC QH541.15.L35 W43 2016 | DDC 577—dc23 LC record available
    at http://lccn.loc.gov/2015036650

This book is dedicated to my parents, Herb and Elaine Weaner, who each in their own way taught me to think for myself. Outside of their love and support, I have never received a more valuable gift. And of course to Team Weaner: Linda, Jan, Sara, and brother Mark. I love you all very much.

—LARRY WEANER

To Suzanne, as always.

—THOMAS CHRISTOPHER

CONTENTS

PREFACE

Every book is born of collaborations between writers, editors, designers, and publishers. This book is even more of a collaboration than most.

When I met Larry Weaner, I was looking for something more, something new. As a horticulturist, I had become keenly aware that the traditional way of doing things in my field was unsustainable, that much of what I was taught to do during my schooldays was actively harmful to the environment. I wanted my garden to be more than just attractive and fruitful. I wanted it to make a positive contribution to the natural scene around it, to design it so that, in its own modest way, my garden would provide what ecologists call "ecosystem services."

When I heard Larry Weaner speak at one of his annual ecological landscaping conferences, I knew I had found my answer. The strategies he laid out promised to make each garden a source of ecological renewal. What's more, he had the experience to prove his strategies could work. By the time I heard him talk, he had been designing gardens for more than three decades and had hundreds of satisfied clients to attest to the effectiveness of his concepts.

Not long afterward, Larry contacted me and suggested that I help him write a book. And so a partnership was born. Relying on my own decades of horticultural experience, I began to pose questions. From the answers, Larry and I began to assemble a text. I contributed my own experiences gained from my work with sustainable lawns. In addition, although Larry is an experienced and expert teacher, my own status as a newcomer to Larry's system of horticulture helped us to identify how to structure the book in a way that was accessible to the average reader.

We chose to tell this story in Larry's voice because it is his story. As I became more involved, however, and began to apply Larry's insights to my own garden, it became my story as well. Our hope is that readers will follow the same path and that this story will become your story, too. As it does, I foresee a time when the collaboration intrinsic to this book will take on a new dimension, and many more gardens and back yards all across North America will make their own contributions to the health and beauty of their local ecosystems. That surely is a collaboration, a new partnership, well worth pursuing.

—*Thomas Christopher*

Neither the hairy alumroot on the right nor the cardinalflower on the left was planted here. Their presence is a result of self-proliferation and human landscape manipulations.

ECOLOGICAL GARDENING
AN INTRODUCTION

I have been profoundly influenced by the activists who, since the publication of Rachel Carson's *Silent Spring* (and in some cases even before that), have been working to reverse the harmful habits into which much of gardening had fallen. There are powerful environmental reasons for bringing our gardens into a sounder relationship with nature. However, I am not hoping to convert readers by teaching them that this kind of gardening is morally or environmentally the right thing to do. Instead, I honestly believe that having once sampled an ecologically driven approach, gardeners won't want to do anything else. For me, the most persuasive reasons are that it's easier and far more rewarding to transform the human landscape in this fashion.

Through my many years of experience, I have learned to treasure the subtle and unpredictable beauties of a garden that collaborates with the local ecology. I prefer to work with a landscape that, at least in part, evolves over time according to natural processes of change and plants itself. I find the diversity of an evolving garden's plant and animal residents fascinating, and the way in which the landscape or garden recruits new inhabitants as it evolves is one of its most satisfying and richest aspects.

A large part of what I advocate is the shedding of conventional gardening's counter-productive practices. The ways in which gardeners till and weed, irrigate, and fertilize their plots, for example, cause perpetual disturbance to the ecologies of those areas and create an irresistible invitation to invasive species. And the ways we were taught to combat those invaders ensure that the struggle will never end. Unfortunately, though perhaps not surprisingly, many of these counter-productive practices have been perpetuated by the more recent schools of "natural gardening" that, though they have an admirable emphasis on the use of native species, still commonly arrange and maintain plantings in a conventional way.

These impediments can best be corrected by a return to first principles, by studying how plants and wildlife associate in a natural state and basing our gardening on that. My experiences have taught me that this change of behavior brings not only better results—a healthier, more dynamic landscape—but also one that demands far fewer inputs. I am selfish enough to have found the latter a powerful motivation to change, and I suspect readers will react similarly.

I don't recall whether this color combination was intentional—but I'll take it.

Cardinalflower and other lowland species were purposefully seeded into a wet depression in this designed meadow.

In this alternative approach, less is truly more. Minimizing intervention and letting the indigenous vegetation dictate plant selection and, as much as possible, do the planting, produces a garden landscape that flourishes without the traditional injections of irrigation and fertilizers and is better able to cope on its own with weeds and pests. The gardener's input becomes a matter of directing the garden ecosystem's evolution into desirable paths and capitalizing on positive developments as they occur. This more fluid style of gardening frees the owner from the drudgery inherent in the traditional attempt to keep the landscape fixed according to some artificial blueprint.

At the same time, this kind of gardening fulfills many of the goals promoted by the environmental activists. It turns the landscape from a consumer of resources and a polluter into a source of environmental renewal: a nexus of stormwater absorption and purification, a sanctuary for indigenous wildlife, and a protector of biodiversity. These are all important, even essential. What I relish most, though, and what I believe will speak most powerfully to readers is the unique beauty and rich experience that this ecologically based style of gardening produces.

For all of those galvanized by the message of such environmental gardening manifestos as Sara Stein's *Noah's Garden* and Doug Tallamy's *Bringing Nature Home*, this book is the next step, the way to turn philosophy into practice.

A NEW KIND OF GARDEN BASED ON AGE-OLD LAWS

I learned to garden and to design gardens the traditional way. What I was taught and what I practiced for some years after graduating college and starting my own garden design business was to create "compositions," "vistas," and "garden rooms." Even when my planting was naturalistic in style, the goal was to reshape the landscape into static, man-made arrangements. That wasn't just *my* goal, of course, it's the goal of any conventional garden designer. And as soon as I or any of these colleagues had installed a design, our goal then

became to perpetuate the original vision more or less intact. Nothing should change the artificial harmonies and contrasts we had contrived.

Maintaining this kind of garden involves a relentless struggle with the site's attempts to express itself. Any unplanned plant that dares to poke up from the soil is a threat to the integrity of the design and must be rooted out, even though it may be better adapted to the local soil and climate than the plants the designer has introduced. Even the species the designer has chosen to include can become enemies if they sow seed and perhaps find a spot they like better than the one they were allotted. Most of the traditional gardener's time is devoted not to making things grow but to keeping things from growing. Ironically, it is the species that prove poorly adapted to which the gardener devotes the most time, because these problem plants must be kept alive—on intensive care if necessary—as letting them go compromises the designer's dictatorial vision.

Eventually, I grew tired of the struggle and recognized that the type of design conventional gardeners practice violates the most fundamental environmental imperative. Nothing is static in nature, every healthy landscape evolves and changes, and its plant community evolves with it at every instant of its existence. In fact, ecologists view change, a gradual evolution, as a basic criterion of ecosystem health. You may admire the complex of wildflowers you find in some natural habitat, but if you return in ten years you will most likely find, if the habitat is functioning normally, that complex changed, with some of the species that dominated during your previous visit declining in number or even vanished altogether and replaced by new arrivals.

When I recognized this, I realized that I must go back and relearn garden design and that, this time, ecology should be my teacher. As I proceeded down this path, I soon came face to face with the fact that it wasn't just my method of design that had to change. So too did the techniques I used to install and maintain my gardens, for traditional horticultural practices are just as divorced from environmental imperatives as the type of gardens they were developed to maintain. An ecologically driven garden clearly required ecologically based techniques.

COUNTER-PRODUCTIVE PRACTICES OF TRADITIONAL HORTICULTURE

As a young horticulturist learning his trade, I had accepted as givens many practices simply because they were labeled as essential by whomever passed them down to me. Later, however, I realized that they are only essential in the context of a traditional kind of garden. When I compared these practices to what I had learned as a garden ecologist, I realized that many of these practices are not only nonessential but instead are actually counter-productive.

Eliminate weeds by pulling them out, roots and all

This may satisfy your need for vengeance, but if the weeds are growing among competitive native plants, disturbing soil in this fashion only creates opportunities for more weed invasion and promotes the germination of dormant weed seeds in the soil. The garden ecologist cuts weeds off at the base, repeating as necessary to prevent them from setting seed and weakening them so that native neighbors can crowd them out.

Fertilize and irrigate to give a boost to established plantings

If you've chosen plants that are adapted to the soil, site, and climate, once rooted in the plantings won't need such artificial assistance (except perhaps in the rare case of an extreme and prolonged drought). Dumping extra and unnecessary resources into a garden ecosystem invites invasion by those garden opportunists, weeds.

The best seeding season is early spring, when temperatures are cool and the weather is moist

Not necessarily. The best season for planting seeds depends on the natural schedule of the species you are planting. For example, native meadow grasses and many meadow flowers, whose natural season of growth comes during warmer weather, benefit from a late spring or even early summer planting. These warm-season plants flourish in summer's hotter weather. The heat of summer discourages the growth of the most aggressive weed invaders, which are non-native plants that prefer cooler, moister weather. A later planting, then, gives the warm-weather meadow grasses and flowers an advantage and a head start over their weedy competitors. Garden ecologists understand and exploit the differences of plants' life cycles to promote desirable species and discourage the undesirables.

Amend soil before planting by deep digging and adding organic matter, topsoil, and fertilizers

The truth is that it's far smarter to plant species that are adapted to the existing character of your soil. This sort of "one size fits all" soil preparation recommended by traditional horticulture is not only labor and resource intensive, it creates a situation that promotes virtually every kind of weed. And the compost, manure, or topsoil you import is likely to bring with it seeds and roots of invasive plants. Typically, a far smaller assortment of weeds is adapted to the more specialized character of your existing soil, which makes their control much easier.

Raise the pH of an acidic soil to make the nutrients it contains more readily available to plants

Raising soil pH also helps to foster weeds, often at the expense of more desirable species. In many cases, hardy native plants are adept at extracting the nutrients they need from acidic soils. Indeed, when establishing native plant meadows, I often deliberately acidify a nutrient-rich soil with a dose of powdered sulfur, which discourages weed growth without adversely affecting the meadow species.

The beauty of the garden depends on prolonging the life of desirable plants by any means available

Trying to keep garden plantings static not only involves a great deal of work, it works against the health and interest of the garden. In the natural landscape, plants are constantly dying, often with new species replacing the old in a developmental process called ecological succession. This sort of flux limits the damage done by plant pests and diseases. If you incorporate this pattern into your gardening, spraying and other forms of disease and pest control become unnecessary. Letting the plant community evolve in this manner also means that the garden never grows boring but is constantly exhibiting new forms of beauty.

A DIFFERENT DESIGN PROCESS

What I have learned amply repays the time I have devoted to study and reinvention. My gardens have become much less laborious and resource intensive and infinitely more interesting than anything I could have created in the traditional controlling manner. What's more, I have the satisfaction of knowing now that when I do my job well, if I base my design choices on an understanding of natural systems and let nature determine the details, my gardens become sustainable in a way a conventional garden never can be.

In fact, a design of the sort I now aim at is more than just sustainable. Because it works with the nature of the site and region, this kind of garden enriches and renews the landscape, raising it to a higher level of ecological function. It increases the benefits that the landscape provides to the surrounding ecosystem. Instead of placing demands on the local water supply, it becomes a contributor. A traditional garden, with its heavy dependence on fertilizer inputs and fossil-fuel-powered maintenance equipment, can be a significant source of greenhouse gases. My type of garden functions as a carbon-sink, removing carbon dioxide (the primary greenhouse gas) from the atmosphere. Rather than displacing the local flora with imported species of little value to the regional wildlife, my gardens serve as sanctuaries for the local animal community and oases of native plant biodiversity.

Getting to this point involved years of study, observation, experimentation, and learning from my failures as well as my successes. My goal is to share this journey with the reader, so that you too can become what I like to describe as a garden ecologist. Your guideposts to design and gardening practice will become what you observe in the landscape itself. You will learn to practice a new, dynamic style of design based on the understanding that because nature is your partner, there can be no precise blueprint of what the garden will look like when is "finished," just a set of parameters beyond which you won't let the garden stray. You will accept that a garden never stops evolving, that it always will be a work in progress. Your role as the gardener will be to watch, interpret, understand, and, at critical moments, give a push to direct the landscape into a path that you can enjoy. The site itself, the natural processes through which it expresses itself, is your design partner.

Let me be clear, though: I am not advocating just letting go. I am a gardener, not just a bystander, and I design my landscapes to suit the tastes and needs of my clients. I accomplish this end, however, by finding a path down which the garden can evolve naturally.

A warning: ecological garden design is not a style of garden-making for the micro-manager. To be successful, this sort of design requires letting the landscape make many of the decisions. The gardener must be alert to hints from the land itself about what sort of plants it naturally tends to support. Weed control isn't eliminated entirely, though. Certain invasive and aggressive plants such as Canada thistle (*Cirsium arvense*), a European native, if allowed to spread unchecked can subvert a newly planted meadow, for instance. But a more relaxed attitude is recommended. Recently, a client for whom I had just planted a 5-acre meadow called to ask what she should do about the dandelions popping up on the grassland-to-be. "Ignore them," I suggested. "They won't be able to compete and will fade away as the taller grasses and wildflowers mature around them." Above all, you cannot declare war on the character of the place and try to force a droughty, alkaline soil to support rhododendrons or camellias or make a humid, lowland garden nurture Mediterranean plants.

NATIVE AND NOT

That is not to say that my style of design must involve only native plants. I prefer to use natives so that my landscapes can integrate with the local ecosystem to provide habitat

A large designed meadow frames a poolside garden. Over time, meadow vegetation has seeded into the garden areas, visually and functionally blending the cultivated and the wild. Increased bird counts have been attributed to this designed landscape, demonstrating that beauty and environmental enhancements need not be mutually exclusive.

for wildlife and a reservoir of genetic material to enrich its surroundings. However, there can be a time and place for exotic species that have proven noninvasive and adapted to the region. Many of us may feel a special affection for some plant of non-native origin, and if it can be included without creating environmental problems, that's fine. Some activities also demand non-native settings: sports such as baseball or soccer, for example, are best played on the sort of Old World turf on which they were invented, which is why a lawn may function more successfully as a children's play space than a meadow of native grasses and flowers.

LETTING THE PLANT FIND ITS NICHE

Success in this style of design requires not only letting the landscape make decisions about what is to be grown there, but also letting the plants find their own niches within the space. Plants must be allowed to explore, to spread their seed around until they find the microclimate and spot that suits them best. An ecologist's definition of a successful plant, after all, is one that not only survives but also proliferates within a habitat or ecosystem. Garden ecologists take their cue from that, learning the ways in which desired species proliferate and using that knowledge to take actions that promote this process. There are also situations, however, when I find myself using an understanding of a plant's ecology to put a brake on its spread. If too successful, it may dominate the landscape in a way that I find unappealing or take the landscape down an evolutionary path that conflicts with the intended use.

INTERVENTIONS

Inevitably, the gardener will need to do a certain amount of editing. A groundcover that favors shade and cool, moist, rock-strewn habitats may be at home in the flagstone terrace where you take refuge during the heat of the summer, but it must be restrained to leave space for you. Likewise, if you garden outside the prairie states and you wish to amplify a view with a sweep of meadow, you will have to intervene from time to time to keep shrubs from invading and initiating the progression from grassland to brush to woodland that is inherent throughout most of eastern North America. But a key to minimizing the inputs of resources and labor required for garden maintenance is to appreciate the flora and patterns that the land produces spontaneously. You may choose to enrich and diversify it, but take the natural growth as your guide. This is why my style of landscape typically requires much less maintenance than the traditional alternative.

PLANTING A RELAY

The example of the natural progression from grassland to woodland that is intrinsic to many regions suggests another fundamental aspect of my ecologically driven, dynamic style of design. Plants occupy niches in time as well as space. Pioneer species, for example, are genetically programmed to thrive in disturbed soils. In the wild, these plants spring up in the wake of a fire, a flood, or a windstorm that strips the land of its plant cover. These pioneers are likely to be a strong presence in a young garden where construction and the initial grooming of the site have caused widespread disturbance, but the pioneers will usually give way over time to other species that are longer lived and more competitive though slower growing. This process of displacement, of one species replacing another, is ordinarily ongoing, although it can be reset by some sort of environmental disturbance that by killing or wiping away the existing vegetation creates a renewed opportunity for pioneers. One way to visualize this process is as a relay, with a team of species that each hands off to the next as it finishes its stint.

This reality makes planning for succession in your planting both a necessity and an opportunity. It's a necessity because if you don't provide successors for the pioneers, undesirable plant species already present on the site or that arrive as wind-blown or animal-transported seed will take over the succession, and they are likely to be aesthetically and ecologically undesirable invasive species such as autumn olive (*Elaeagnus umbellata*) or Japanese honeysuckle (*Lonicera japonica*). Planting a relay is also an opportunity because it allows you to set the direction in which the landscape will evolve, at least to some extent.

REWARDS

There are many rewards to gardening in a more ecologically informed style. I have already mentioned the reduced need for maintenance. Because such a landscape, once set in motion, largely plants itself, the budget for new planting should decrease greatly over time. This is in marked contrast to gardens of a more traditional type, where an artificial plan forces plants to inhabit spots whose conditions may not suit them, so that replacement becomes routine. In addition, whereas an ecological garden may be allowed to expand by self-seeding or the natural spread of existing root systems if the space is available, expansion of a conventional garden requires the purchase of a whole new set of plants. Finally, because the plants in an ecological garden have been selected for their adaption to the existing environmental conditions, once they are rooted in they should not require irrigation or fertilization.

I've mentioned the benefits to the local ecosystem and wildlife such a garden can provide. There's another reward as well, and that is this: no matter how long you live with it, an ecologically designed landscape never becomes routine. My own garden, a ⅓-acre suburban lot outside of Philadelphia, has never stopped evolving over the twenty-eight years in which I've tended it. I have come to know that patch intimately as a result of my gardening there, and yet it continues to surprise me year after year; there is always something new as it continues to seek its own course. This is one of the great advantages of ecological garden design. Designers often speak of the need to instill a sense of mystery into the garden: just as romance dies when a relationship becomes too predictable, so too a predictable garden rapidly becomes boring. The changes I find in my garden from day to day may be subtle, but they are rarely predictable.

Plants shown here were grown from seed without fertilizer, irrigation, or extensive weeding.

THE
LEARNING
PROCESS

THE RISE AND FALL (AND RISE) OF A CARDINALFLOWER PATCH

It was midsummer and the end of a long horseback ride through a swampy wood in Cape May, New Jersey, when suddenly I came upon the patch of cardinalflower (*Lobelia cardinalis*) ablaze in full scarlet bloom. I hadn't seen a single specimen of this native wildflower anywhere else in the wood, though presumably the moist, organic-rich soil provided suitable habitat nearly throughout. Why did the plant flourish only in this one spot?

A natural presumption would be that the cardinalflower bloomed only here because this was the only spot where its seed had somehow been deposited. This, however, I knew to be an unlikely scenario, given that cardinalflower is a common inhabitant of such wetlands throughout most of the eastern, central, and southern United States. A closer inspection of the site suggested another explanation, one with important implications for my design work.

I noticed a fallen tree, at which point it all began to make sense. As this tree crashed to the ground, most likely blown down by a storm, its fall ripped open a gap in the woodland's canopy of branches and foliage, letting sun reach the woodland floor. When the falling tree's branches hit the ground, they dislodged the existing groundcover of plants and stirred the soil. This event, known as a disturbance in ecological terminology, provided an opportunity for cardinalflower seeds that had lain dormant in the soil, possibly for many years, waiting for the exposure and sunlight that would trigger germination and support the plants' subsequent growth. Almost certainly, there were dormant cardinalflower seeds throughout the swamp. They had germinated only in this spot because of the tree's fall.

If I were to return to the same spot today I would probably not find cardinalflowers growing there. More likely I would find in the ground layer of vegetation the kinds of plants that had flourished on the site before the tree's fall. The cardinalflowers had served as a pioneer species, with their rapid growth enabling them to fulfill the ecological purpose of stabilizing the disturbed soil. With time, though, stronger and long-lived competitors, their predecessors in the spot, would reassert themselves, crowding out the cardinalflowers so that once again all that remained of the wildflower were dormant seeds of the most recent generation, waiting for the next disturbance, which might come in ten years or perhaps a hundred.

The horticultural guides that a gardener might consult state that cardinalflower is a short-lived perennial that seeds itself readily throughout the garden, both of which are qualities that are problematic in a traditional, static design. But an understanding of the species' ecological role enables the gardener to understand how these qualities can be useful even in a cultivated landscape.

In later years I planted cardinalflower and blue flag iris (*Iris versicolor*) in a newly dug drainage swale in a garden I was creating in a residential neighborhood. The digging of the swale was another disturbance event, although in this case a bulldozer played the role of the falling tree. I planted the cardinalflower not just to please the client's eye and attract hummingbirds to the

Unless the soil is disturbed, this patch of ground will soon be owned by the blue-flowered tall phlox, as it is more long lived and competitive than its ephemeral colleagues, blackeyed Susan and cardinalflower.

top If you plant cardinalflower, you will see hummingbirds.

bottom The red spikes of cardinalflower emerging near a felled tree in a New Jersey woodland prompted some of my first realizations about how landscapes change and, in turn, the role we can play in these changes.

landscape, but also to serve the same function as it did in the swamp: to germinate quickly, stabilize the soil, and provide cover while the slower growing but ultimately more competitive iris became established.

By planting these two species together, I replicated the process that takes place after a disturbance in the wild. These two wetland wildflowers are working in a complementary fashion over time to provide short- and long-term stabilization of the soil disturbed by the excavation.

There's another chapter in this story, and it demonstrates the value of observing a place over time and thinking analytically about what has happened there. Six years later, I returned to the garden and found that the cardinalflower was still flourishing and hadn't been displaced by the iris. Why, on this site, had the cardinalflower persisted beyond the limited time period that its ecological role would suggest? I checked the soil below the plants, and I found in it small furrows cut by rushing water funneled through the swale by storms. In effect, the runoff was causing frequent soil disturbance, creating openings for continual germination of cardinalflower seeds and renewal of the plant's life cycle.

What value can this observation have for a gardener? Cardinalflower is a beautiful plant, but one whose limited duration in the landscape disappoints many gardeners. I had observed a condition under which the plant became a permanent player in the landscape. Why not replicate this in a garden setting? To accomplish this, I planted a combination of bluestar (*Amsonia hubrichtii*), tall phlox (*Phlox paniculata*), and cardinalflower. The bluestar and phlox are competitive and long-lived native perennials that I selected not only for their showy flowers but also because they form dense colonies that suppress the germination of weed seeds.

Cardinalflower envelops a seating area in this wooded garden.

Ordinarily, the short-lived cardinalflower would fade from this planting relatively quickly. However, here my advice helped the homeowner to perpetuate it by replicating the process demonstrated in the swale and the Cape May swamp. After the seed has fallen, the homeowner lightly scratches the soil immediately below each cardinalflower. Essentially, a rake performs the same function as the rushing water does in the swale or the falling tree did in the swamp. The gardener is using pin-pointed disturbance to promote germination of the cardinalflower seeds without upsetting the stable and weed-prohibitive cover maintained around it by the other more competitive perennials. In this way, a natural process has been incorporated into the garden's maintenance routine, so that a couple of minutes of easy labor eliminates the need for repeated replanting that traditional horticultural practice would dictate.

Fifteen years had elapsed between the horseback ride in the swamp and my creation of this garden. Just as the population of plants in a woodland evolved over time, so too had my understanding of the natural environment. And that has been so only because I paid attention to the people and places that had something to say.

INFLUENCES AND INNOVATORS

If you've ever taken the train from New York to Washington, D.C., while passing through Philadelphia you saw the house I grew up in: two stories and a postage-stamp back yard tucked in against the railroad tracks. Many of my gardening friends have told me how they learned to love their craft planting, weeding, or watering alongside their parents. I didn't. My mother had no interest in outdoor hobbies; my father planted one dogwood tree in the front yard and spent an hour every weekend during the spring and summer cutting our modest patch of grass. And that, as far as he was concerned, was that. Even later, when I became interested in landscape design and used our yard to create a garden of rare and interesting plants, my father had only one question whenever I installed something new: "That thing isn't gonna get too big, is it?"

Clearly, I can't claim gardening as a heritage, but in some respects I think that's a good thing. I wasn't indoctrinated with traditional gardening principles when I was young and impressionable. My parents never taught me to regard double-digging as the apotheosis of the craft. When I was finally introduced to that painfully laborious process (and all the other back-killing techniques revered by traditional gardeners) I was old enough to think for myself. Also, long before then I had already formed the view of nature that still informs my work. Because I wasn't first introduced to plants as individuals in a garden setting, but instead encountered them as part of wild communities, I've never really shared traditional gardening's preoccupation with specimens or even individual plants per se. Mostly it was the place that attracted me—a place to stumble upon, to admire, and to figure out. I interact with such a place but do not feel the urge to convert it into a mere reflection of my own tastes. For me, a garden at its best is like nature, a place to get lost.

With an astute eye for a site's natural features, the noted landscape architect A. E. Bye understood how to subtly place plants and stone such that the resulting landscape appeared to visitors not to have been designed at all.

PLACES

Places of mystery need not be remote or pristine, a condition impossible to find in the ecosystem that is urban Philadelphia. Consequently, my early explorations in nature consisted of neglected waste places where only the toughest plants survived, forming a wild and tumultuous tangle. Even before I was old enough to cross the street, I would slip beyond our back yard to explore the adjacent railroad right-of-way. (Please don't tell my mom.) That may not seem like much of a wilderness to anyone raised in the country or even a suburb, but to me it sure beat the lawn and that lonely dogwood. And, in fact, such linear open spaces are important corridors for a surprising variety of fauna and flora including birds, insects, wildflowers, and grasses. In the Midwest, for example, railroad rights-of-way preserve some of the last remnants of the native prairies. The right-of-way behind my house was not such a natural space; aside from the gravel railroad bed, it was for the most part a blanket of invasive vines. But it wasn't the plants that absorbed me so much as the quest to find someplace unfamiliar and new—new to me, at least. Clearly, there had been previous explorers down this stretch of earth. Once, I even found an arrowhead there.

Later, when I was old enough to explore outside my immediate neighborhood, I extended my expeditions to nearby Burholme Park. This 70-acre former private estate offered, along with ball fields, picnic grounds, and an eerie museum in the former family mansion, a small scraggly woodland remnant. In retrospect, I realize that nearly all the undergrowth there consisted of invasive species and the creek that bisected the woods was modest at best—it disappeared abruptly into a culvert under a busy street along the park's boundary.

Nevertheless this place was wider than my railroad track corridor. More space to explore, more nooks and crannies to discover, and more potential hideouts. Even after I had trekked down every path, I'd stop, look at a particular slope or dip, and wonder if that hid a scene I hadn't observed before. I'd be looking under rocks, walking up the little creek until I knew every corner of the place and was ready to move on.

This began a pattern I have continued to follow, of stepping off the habitual trail. In doing so, I have found scenes that remain fresh in my memory, shaping the way I may treat a landscape feature. When I was later living in Massachusetts, there was a hike I used to take in a small state forest. Now I was in a real forest, where more space and less human disturbance allowed for a more healthy and graceful landscape, one with intricacy and detail worth looking at closely. Here I started learning to identify trees and woodland wildflowers and to pay attention to the intricate patterns expressed by the plants in their wild but balanced environment.

While bushwhacking off the trail, I found a secluded rock bald from whose top there was a particularly fine view of the surrounding woodland. It covered a large area, and the dramatic interplay of plants and stones at that spot has guided my hand often while designing stone terraces and associated plantings.

This impulse to step out beyond the borders of my control, to get lost, to find some seemingly unvisited, lost place has never left me. When I was nineteen, I set out with a friend to backpack around South America; we parted company when I insisted on walking into the jungle in Ecuador. Travelling by myself, I reached the end of the road, literally, in a frontier town. There I met another American traveler, and together we set off down a jungle trail that supposedly would take us, in about four hours, to a tribal village. After four

days of walking and sleeping in our hammocks, we reached the village. The inhabitants spoke no Spanish, only their own Quechuan dialect, a relic of the time when this remote jungle had been part of the Incan empire. The villagers lived in thatched huts along a river. I was fascinated by this place and its people, but when I found the small chapel a pair of nuns had built there, I wondered what it would be like to journey further down the river to a place where even this small vestige of Western culture hadn't penetrated.

This looking for the unknown and unanticipated is, for me, an essential attraction of ecological gardening. By letting the landscape and the native plant and animal community make many of the design decisions, I return spontaneity and a sense of discovery to the garden.

How can we attain this sense of discovery on a limited property where there is nothing left to explore spatially? One answer is to introduce the element of time by using a design approach based on ecological processes to introduce and manage plant compositions that evolve and change, as plant communities almost always do in nature.

Red columbine seeded itself around the small front garden of the Philadelphia house where I grew up. In those days, I avoided columbine because of its tendency to seed around the landscape and mess up my arrangement. Times have changed.

Visits to a rock bald early in my career inspired this expansive terrace and surrounding gardens, part of an overall landscape plan I developed for a large estate.

An example of the impact such design can have is the house in a Philadelphia suburb where I have lived since 1988. It didn't take long from when my family and I moved in to complete an exploration of the ⅓-acre yard that surrounded the house. This was a fairly standard example of suburban landscaping when we arrived: all lawn, the evergreen groundcovers pachysandra and English ivy (*Hedera helix*), and a handful of overpruned shrubs. The only exceptional features were the trees: a massive scarlet oak (*Quercus coccinea*) in the front yard, a huge white oak (*Quercus alba*) in the back yard, and a big shagbark hickory (*Carya ovata*) just over the fence in a neighbor's yard.

My initial planting was largely ad hoc: I was too busy with clients' gardens to work in a deliberate fashion on my own place. I would introduce plants here and there. Some were thoughtfully selected, and others were merely left over from clients' projects. I did intend to create a plan—next month. But next month never came, and in the meantime nature was creating its own plan. Some of my leftover additions lived, some died, some proliferated, and some new things showed up on their own. The site was essentially indicating what type of vegetation it would support.

My job slowly transferred from planter to editor. The one joe pye weed (*Eutrochium* species) I planted produced seedlings. Those that sprang up along the property perimeter I let grow; those whose 7-foot-tall mature height would block the view from my window I removed. The invasive Oriental bittersweet vine (*Celastrus orbiculatus*) that emerged was pulled out, whereas the native sassafras (*Sassafras albidum*) sapling that sprouted next to it was left to colonize. The cover and food sources this new flora provided attracted more birds, which deposited more seeds. The oak (*Quercus* species) and hickory (*Carya* species) trees that chipmunks had planted were no longer being mowed down because the lawn I'd inherited from the previous owner had been colonized and crowded out by moss and other native plants.

I had managed this landscape. Surely, if I had done nothing it would have become a weedy mess. But the activities I did perform, while physically minimal, were thoughtfully formulated to favor the species I wanted over those that I didn't. Essentially, I was favoring brains over brawn. My motivation for all of this was twofold: a desire to observe and participate in the ecological processes that were unfolding on my property, and a desire to perform as little yard work as possible.

I no longer needed more space to encounter unexpected surprises in my landscape. They were now presented to me through the passage of time.

JOBS

My initial introduction to landscape design came with a job I took during the summer after graduating from high school. I answered an ad in the newspaper, and, after a trial day, I was hired by Eugene Varady, a Philadelphia landscaper. Varady liked to test two applicants for a position on the same day and pick only one. I, a skinny young kid, was pitted against a big, muscular Navy vet. Varady was ex-Navy himself: "two sailors and a Hail Mary" was how he characterized the crew that day.

Until I saw Varady's ad, I hadn't been aware such a profession as landscape designer existed. In my neighborhood, if you wanted a new plant for your yard, you went to a garden center and plunked the purchase into the ground yourself.

Varady was a tempestuous man who had difficulty getting along with employees and customers, but he was very creative in his design, even though he worked on a small scale. He didn't work from a paper plan. Instead, he would select plants for a project and arrange them on site. Varady's knowledge of plants was deep. I liked the passionate and improvisational manner with which he approached his work, and I found his habit of questioning assumptions, horticultural and otherwise, quite stimulating. I suspect that this shared attitude was the reason that I, not the muscular veteran, got the job.

I ended up working for Varady through several gardening seasons. He'd lay me off at the end of every autumn, and I would travel during the winters, returning to Philadelphia when the planting season began in the spring. The environmental movement in the United States was relatively young then, and one might assume that landscaping was an ecologically beneficial activity. One day, however, Varady posed a question: Was what we were doing good or bad? Yes and no, was our conclusion, when we compared our planting activities with the then nearly ubiquitous use of pesticides, soil erosion as a result of tilling, and other problems.

It is interesting that my first introduction to landscape design was with an improvising iconoclast who constantly questioned assumptions. Hopefully my skills in the personal relations realm are a bit smoother than Varady's, but the road on which I eventually embarked was clearly influenced by his highly creative and individualistic approach.

After working for Eugene Varady for several years and returning to college to study horticulture, I landed a job at Weston Nurseries in Hopkinton, Massachusetts. During my five years there, I became steeped in the world of rare and unusual plants, native and otherwise.

At that time, Weston offered one of the broadest selections of plant species and cultivars (many of which it had bred) to be found in any American nursery. Its stock included many natives, especially North American azaleas and rhododendrons, plants that could be found for sale almost nowhere else at that time. Ed Mezitt, son of the nursery's founders, had passed on the daily operations to the next generation so that he could focus on the development of new plants. His interest in selecting and hybridizing native plants was very innovative for the time. He would travel to Appalachia periodically to collect rhododendron seeds from the wild. By crossing one of these natives with an Asian rhododendron, he had created the famous PJM group of hybrids that are still ubiquitous throughout New

England. Mezitt also bred several lesser-known hybrids derived from native azalea species, including *Rhododendron* 'Pennsylvania', one that blooms into September, long after the flowering season of the Asian azaleas popular in the trade has ended.

During this period, the mid-1970s, native plants were often viewed as scraggly weeds with small white flowers. "Why would you want to spend money on a plant that grows on its own in the wild?" one nurseryman asked me at the time. This is actually a good question.

I learned an incredible amount about plants during my time at Weston, and while there I strived to become a serious plant maven. I later realized, however, that even then I was looking at plants from a different perspective. I appreciated them for what they could do for me as a designer, rather than for their individual ornamental characteristics. As a place more than a plant person, I still view plants similarly, but now my perspective has been broadened to include not only their potential contribution to my vision, but also to nature's.

In the autumn of 1981, I decided to leave Weston Nurseries to start my own landscape design and build firm the following spring. In the intervening winter I took a trip to Africa with another Weston refugee, and I was fortunate to receive a minor commission from the Brooklyn Botanic Garden to collect and bring back seeds for their medicinal plants collection. This delighted me, not because there was any significant money involved, but because I would to be

In 1982, when I started my business, I had a small operation: myself and one helper.

able to play plant explorer like my first horticultural hero, the great E. H. Wilson (more commonly known as "Chinese" Wilson because of his many successful expeditions to that country). The operative word however was *play*, as I soon realized that my botanical abilities were nowhere near up to snuff. I did return with some seeds; however, they all came from a government of Kenya seed bank facility, which could have been obtained quite easily through a simple phone request. The Brooklyn Botanic Garden did use the seeds, but I declined any payment as my plant hunting consisted of one Nairobi cab ride.

In the spring of 1982, I founded Larry Weaner Landscape Associates in the Berkshire Hills of western Massachusetts, and five years later I relocated to Philadelphia, my hometown and current base. Often through trial and error, I developed the cause and effect, brains over brawn, theoretical versus real-world concepts I describe through a long process of observation and analysis put into practice. In the end, I learned that a traditional garden is like a beautiful car with no engine. The body is sleek, the interior is plush, and the stereo sounds great, but the owner will always need to push it up the hills with bags of fertilizer, weeding forks, and watering wands.

MENTORS

In my early days with Eugene Varady and at Weston Nurseries, I considered my work to be naturalistic, as did many designers and gardeners at that time. Though this style encouraged informal plant arrangements over straight lines and clipped hedges, it was not typically based on the observation of wild plant compositions, nor was the use of native plants a main focus for most practitioners. Incorporating natural processes of change into the design process was not even a glint in the designer's eye. Truthfully, naturalistic really just meant informal in most cases.

The first influence that guided me from naturalistic design to a more literal approach came during a three-day course on native meadows I took at the Harvard Graduate School of Design in 1984 with the renowned landscape architects A. E. Bye and Ted Browning. I had already planted a few meadows for clients, but, not having any background in ecology at the time, what I planted had proven short lived.

Bye and Browning made me realize that nature is highly structured and that things happen in the natural landscape for a reason. A meadow isn't just a place with pretty plants and no straight lines. Its patterns, processes, and interactions make it what it is. In the course of a few days with these two masters, I learned that if you want to plant a meadow, it's not going to work unless you understand the underlying ecology of an open field or grassland. I learned, for example, that a grass is not just a grass: there are warm-season clump-forming grasses and cool-season sod-forming grasses, and only if you understand their differing life cycles and growth patterns can you influence which type will dominate. Given that warm-season clump-forming grasses are largely compatible with native wildflowers and cool-season sod formers crowd them out, this is kind of a big deal. I learned that there are a lot of big deals in the ecology of an open field, and that similar discoveries could be made and used to enhance the effectiveness of planting in woodlands and other habitats.

During these eye-opening sessions, Ted Browning drove home the point that nature is ordered, that even if it isn't entirely predictable there are known (or at least knowable) forces and processes governing the behavior of plants, all of which is key to creating a sustainable meadow. When I design now, regardless of the landscape type, I seek to work within that order.

Bye's yin to Browning's yang was his highly visual approach, which contrasted significantly with Browning's more science-based analysis. Bye spent a lifetime intensely observing nature, from expansive scenes to intimate plant vignettes, and was able to create gardens that captured the visual essence of these landscapes in extraordinary detail and accuracy. Instead of me waxing poetic and vainly trying to capture the essence of Bye capturing the essence of nature, I would suggest you just look at the image of a Connecticut woodland landscape created and photographed by Bye himself.

In essence, these two innovative designers taught me that in order to create gardens that reflect nature, you need to understand and absorb nature: the plants that you see, as well as the underlying processes that you don't.

Once you begin to understand ecological processes, what do you with them? This question was answered for me by the brilliant and innovative ecologist Frank Egler.

Frank Egler planted *Liatris* species and tracked their slow but steady proliferation over time.

Canada goldenrod, a highly aggressive native wildflower, can form monocultures in fields with rich soils. Egler noted that it did not dominate in this field, even after forty years, and he theorized that the poor gravelly soil limited the species.

left Egler experimented with mowing times and found that mowing a path in June favored little bluestem, a warm-season grass still emerging in late spring.

opposite page Alternatively, mowing a path in August favored cool-season grasses over the little bluestem.

Though Egler had degrees in ecology from the University of Chicago, the University of Minnesota, and Yale, he was too outspoken and argumentative to keep an academic position. In his mid-thirties, he retired to his family's 500-acre country estate in Norfolk, Connecticut, where he experimented with techniques for managing wild and cultivated landscapes based on natural ecological processes. He kept detailed notes on all of his work there, so that he could deduce the cause and effect of each technique he employed. Because he continued his work on the site for more than fifty years, his results offer a uniquely long-term perspective on managed vegetative change, a field he was essentially inventing as he went.

I learned about Egler from William Niering and Glenn Dreyer, executive directors of the Connecticut College Arboretum, who encouraged me to read his publications and visit his property, Aton Forest, which was put in a conservancy to continue Egler's work after his death, when John Anderson took over as executive director.

A notable example of Egler's radical new take on land management is a 7-acre field, known as Woodchuck Hill. Today, this is almost entirely covered by low shrubs such as low-bush blueberry (*Vaccinium angustifolium*), sweet fern (*Comptonia peregrina*), and spreading junipers (*Juniperus* species). When Egler retired to this property in the 1940s, this particular piece of land was a hayfield. He decided to stop mowing the field to let it develop into something more natural. When he found stunted specimens of shrubs among the grasses, Egler realized that this was the direction in which this field would evolve, which made sense, given that the soil was dry and thin with rock ledge near the surface, ideal habitat for these shrubs.

In fact, when Egler stopped mowing, the shrubs did begin to overwhelm the grass. At the same time, however, trees also sprang up, threatening to return the field to woodland. To keep the field an open shrubland, he cut down the trees while they were still young and painted their stumps with herbicide. (It should be noted that Egler was the first person to use herbicides for ecological restoration. He spot applied but never broadcast the materials and was described in Rachel Carson's *Silent Spring* as one of the few people who used herbicides responsibly.) The shrubs reacted favorably to this regime, while the number of new tree saplings springing up gradually declined.

Seeking the cause-and-effect relationship in this and subsequent experiments, Egler began to develop his theory of initial floristic composition. It was a tenet of ecology at that time that land that is disturbed in some way, such as by being cleared to make a hayfield, will experience succession once the disturbance ceases. Thanks to seeds carried in by birds, wind, or other natural factors, the vegetation on the site would spontaneously evolve toward whatever was the climax vegetation for that region (in northwestern Connecticut, that would in most cases be forest). Egler asserted that, on the contrary, the majority of the plants that grew on any disturbed land sprang, at least in the first several decades, principally from seeds already on the site when the disturbance occurred. Once vegetation became dense cover it would suppress the germination of most new seeds entering the system.

Frank Egler spent fifty years establishing this shrubland on Woodchuck Hill by removing trees. Here, ecologist John Anderson discusses Egler's work at a landscape workshop.

The dynamic that Egler observed and described was the basis for his accomplishment in directing the succession of the vegetation on Woodchuck Hill. The numerous trees that emerged when Egler stopped mowing were not new arrivals but had been present (albeit suppressed) when the tract was still a hayfield. When the mowing stopped, the trees were released—enabled to grow freely—and sprang up in abundance. Egler posited that if he removed the trees that had already been present on the site, the shrubs that were also present would have a head start on any new tree seeds that found their way into the ecosystem. The dense cover the shrubs would form would greatly reduce the germination of those newly arrived tree seeds, so that the shrubland would tend to dominate for a long period of time.

The implications of this theory for the land manager is that after determining what sorts of plants a site is predisposed to support, it is most important to begin active management and species editing very early in the process. If, for example, you are creating a

meadow, focus on removing weeds early in the development of the area so that the perennial meadow plants can knit together and bar incursions from the outside. Gardeners can become discouraged by the intensity of the care required in the early stages of realizing an ecological garden design. But Egler's theory shows that work done then is more likely to create a highly stable and more easily managed community down the road.

I've taken many to view the shrubland at Woodchuck Hill. Surrounding the field is a forest of 60-foot-tall trees. This tract was part of the original hayfield and was designated as the control in Egler's experiment. Consequently it was never managed. Visitors initially think that the shrub-dominated field is naturally occurring, but in actuality it is the surrounding woodland that is a natural occurrence. The shrub carpet was created, or maybe a better word is fostered, by Egler. Essentially he prevented the onset of a 60-foot-tall forest

The landscape at Aton Forest illustrates how much can be learned by observing ecological change in one place over a long period. Ecology aside, Aton Forest is quite beautiful in its evolving intricacy, a testament to the fact that scientists sometimes have a lot to teach designers about design.

Using methods that we're only just beginning to understand and appreciate, Native Americans manipulated landscapes like this one in Yosemite at both minute and vast scales to meet their needs for food, clothing, and shelter.

on 7 acres over a fifty-year period by himself, without ever mowing, without employing a weeding crew, and without using any power equipment. How did he do this? With an approach that was familiar to me: brains over brawn.

Egler's promotion of initial floristic composition and his application of it at Woodchuck Hill showed me that ecological theory could be more than an academic exercise, but could suggest practical strategies to make native plant establishment easier, more effective, and at times even possible.

The next game-changing idea for me—one that has only come to gardeners and designers recently and is still controversial in some circles—is that humanity and nature need not be at odds and in fact are inseparable. In the past, many preservers and restorers of natural areas believed that fundamental to their task was eliminating any evidence of human activity. I came to view this process quite differently after visiting Yosemite National Park, where I began reading *Tending the Wild*, M. Kat Anderson's book about

Native American land management practices in California prior to European colonization. Anderson is an enthnoecologist who has served as a consultant to the National Park Service, the U.S. Forest Service, and the California Department of Parks and Recreation and as a lecturer at the University of California, Davis, in the Department of Environmental Horticulture.

In *Tending the Wild*, Anderson confronts the somewhat romantic vision of this landscape popularized by the founder of the Sierra Club and the man most responsible for Yosemite's preservation, John Muir. Despite his remarkable insights as a self-taught geologist and botanist, Muir fundamentally misunderstood the ecological history of the California wilderness that he loved. He believed that it had been pristine and untouched before the arrival of Europeans. In fact, as Anderson revealed by her painstaking study of Native American interactions with the landscape—gained from many sources, including accounts of early Euro-American explorers and recollections by descendants of the Native inhabitants—the sweeping grassland and wildflower vistas that Muir admired in Yosemite were largely an artifact of Native American management activities. For thousands of years, the tribes of the Sierra Nevada region practiced seed dispersal, transplanting, pruning, and above all controlled burning over vast areas to favor the plants and landscape types that would best sustain them.

Furthermore, Native Americans not only applied these techniques to the landscapes of Yosemite but to the entire state of California. Their effect was so dramatic that the seemingly wild landscapes encountered by the first European explorers were in large part anthropogenic creations that would likely have appeared quite different had the Natives never been present.

How could a people with no modern equipment or technology effectively manage a landscape as vast as the state of California, or most of North America as is now posited by some ethnoecologists? The answer once again is brains over brawn. Experiencing the landscape intimately on a daily basis and observing the results of their actions over centuries likely allowed them to accumulate a far greater working knowledge of restoration ecology than any practitioner working today.

Even though these activities were performed over vast areas, they were designed to avoid depletion of the resources on which their livelihoods depended. One example is the lasting effect that Native American management had on Santa Barbara sedge (*Carex barbarae*), which covered broad meadows at the time of the first descriptions written by Euro-American visitors. The Native inhabitants had harvested this plant, using the foliage for basket making and eating the roots. It was their practice to split each clump of sedge they dug and return half to the ground. This technique ensured their long-term access to the resource, a restorative approach that modern land managers have only recently begun to grasp.

Beyond their activist yet preservationist approach to managing vegetation, the Native Americans and their sedge meadows can teach us another valuable lesson about the effects of long-term interactions between plants and people. According to Anderson, when the Native Americans were displaced from these areas and ceased harvesting in them, the sedge meadows began to deteriorate. Evidently, over centuries of human use, the plant had become dependent on regular digging and replanting for renewal and even survival. When we try to re-establish native species in our landscapes or wild places, understanding the historic management practices on which they still may be genetically dependent could prove very useful or even essential.

A clear example of how this kind of knowledge could be useful emerges from accounts of the Californian tribes' use of precision burning around the oaks whose acorns they depended on for food. The area beneath the oak canopies harbored insects that laid eggs in the ripening acorns, destroying the acorns' value as food and even their viability as seed. The Natives had learned, however, that burning the grass or underbrush beneath the oak canopies in autumn drastically reduced the insect populations and ensured a better harvest from those particular trees the following year.

In modern times few people eat acorns, so why should we care what the Native Americans did? We do plant oaks, and we try to restore oak forests for ecological and aesthetic reasons. This ancient management technique could perhaps be very useful, protecting seed crops without the use of the kind of toxic sprays on which modern land managers rely.

In fact, in recent years the National Park Service has also adopted prescribed fires—controlled burns of the landscape similar to the Native model—as an essential part of its land management techniques. The slogan that the Park Service has promoted for decades with its visitors, "Leave nothing but footprints," still applies to trash and misuse of facilities, but it no longer applies automatically to land management.

Yosemite was made a national park in 1890, and several years later the Park Service brought back some of the elder tribal members who had lived there to comment on what they saw. According to Anderson, a common remark was "This land is unkempt. No one is tending it."

To me, this challenged the idea that natural gardens should be based literally on a time before evil white people perpetrated the first human disturbance on the land. The Native Americans had been disturbing the land for centuries, most likely in a more thoughtful and constructive way, but disturbing it nonetheless. We didn't need a throwback approach but a throw forward one: an approach that builds on historic practices and adds creative new ones that accommodate the ecological and cultural realities of the contemporary landscape.

CONVERSATIONS

As Frank Egler wrote in 1977, "Nature is not more complex than we think, but more complex than we can think." Clearly no single person can know it all. Conversations with others, be they professional colleagues, informal teachers, or ecologically minded neighbors, can help solve some of the ecological riddles that analyzing landscape vegetation processes can generate.

One such conversation occurred in my office several years ago. I was discussing with Ian Caton, one of our designers, the prospect of establishing a woodland wildflower ground

Ian Caton, photographed in a
pond we designed, possesses
phenomenal knowledge
of native plants and their
natural habitats, informed
by his many hours studying
plants in the wild.

layer from seed in a project we were developing. Ian had created a seed mix that included spotted geranium (*Geranium maculatum*). I had planted spotted geranium from seed on projects previously with absolutely no success, and I pointed out that in order to get this plant to colonize the landscape it would have to be done with live plants. Ian's response was that he had put a few live plants in his garden at home, and they had seeded them- selves prolifically throughout the landscape. Why did he get plants from seed and I did not? We both knew that seeds from woodland species often have a particularly short period of viability. As we discussed the question further we realized that by the time the seed I had ordered was harvested, stored at the seed house while waiting to be sold, shipped to me, and eventually planted, it likely was no longer viable. In contrast, Ian's seeds were derived from the plants he had placed directly on the site. Instead of passing through multiple storage and shipping processes, these seeds were hitting the ground at the exact moment they were ripe. Their seeding time was occurring exactly according to plan; not my plan, but nature's. Out of this conversation we developed a woodland wildflower establishment protocol in which mother colonies of live plants were planted in patches throughout the landscape to provide a living seed source for colonization over time.

At the time of this conversation I had been planting woodland landscapes for many years, and Ian's knowledge of native plants was and is so extensive that I often refer to him as a plant genius. Yet the solution to the geranium dilemma was only discovered through an interchange of ideas that drew from both of our experiences.

If two heads are better than one, then ten, twenty, or a hundred heads are better than two. With that in mind I developed a natural design conference series in 1990 for

professional practitioners titled New Directions in the American Landscape. For two days every January we assemble a group of highly knowledgeable presenters from an array of related disciplines. In addition to horticulture and design, presenters are drawn from the worlds of ecology, botany, history, anthropology, and art, among others. There is no basic textbook for ecological gardening, and I realized that in order to create the landscapes we were advocating, landscapes that drew simultaneously from nature, culture, and art, we would have to reach beyond the fields of horticulture and garden design.

As I expected, these diverse groups of presenters bring myriad new perspectives to the table and the response has been enthusiastic. What I did not expect, however, was the conversational synergy that occurs between conference participants as they bat around the concepts that the speakers present.

One presenter was the renowned landscape architect and keen observer of nature A. E. Bye. I was lucky enough to spend time with Bye as I drove him to the conference. During the entire ride, Bye had his head halfway out the window looking at passing landscapes. Even in the middle of a conversation, he was looking—up into trees, out across fields—while he muttered comments upon what he was noticing. I realized that Bye was storing these observations and images in his memory, to be taken out later for consideration and interpretation. As we were not really driving through anything particularly special, I realized that he must do this all of the time. No wonder he could create natural landscapes with such precise visual integrity.

I've picked up Bye's habit of automobile botanizing. (I don't text while driving, but this is probably worse.) In a busy life, I have less time to wander the countryside than I would like. The hours I spend behind the wheel have become precious downtime that I use to absorb the macro themes of the landscape. In particular, I find road cuts, where profiles of the underlying soil and rock are revealed, intriguing. What I've observed again and again is that healthy communities of native plants are much more likely to occur on these exposed rock outcroppings where the soil is thin and water and nutrients are scarce. Low-lying terrains, the bottomlands, where soils are deeper and moisture more abundant are much more likely to be overgrown with tangles of invasive species. This has led to one of the many counter-intuitive realizations I've derived from observing natural areas: for anyone intent on growing native plants, particularly on a scale that you can't manually weed, deep and moisture-retentive soil is not necessarily a blessing.

This leads me to the primary reason I wrote this book. I believe we can create landscapes that are easier to manage, more ecologically beneficial, and reflective of the beautiful plant compositions of our respective regions only if we expand our studies beyond the confines of traditional garden design. In fact, when it comes to creating landscapes that are truly based on nature, much of what we learned in horticulture and design plays out like the old George Gershwin tune, "It Ain't Necessarily So." I have spent more than thirty years absorbing the influences described in these pages and using them to help me make sense of nature's process. My hope is that sharing these experiences will help shorten your educational journey.

Plants like spotted geranium, found throughout eastern North America, can self-seed readily. But as with many woodland wildflowers, its seed does not remain viable for long.

THE MAKING OF A GARDEN STAR

The Garden Conservancy has taken as its mission to identify exceptional gardens throughout the United States, and acts as a catalyst in raising money and public awareness to restore and preserve these masterpieces. Knowing this, I was overjoyed when I learned that the Garden Conservancy's Society of Fellows, its top-level supporters, would be making a special visit to a property in northwestern Connecticut where I had planted and still managed many acres of meadows and woodlands.

The timing of the visit worried me, though. The tour was to take place in October, by which time the floral display in the 40 acres of meadows I had planted would be, for the most part, past. Personally, I love this meadow's appearance in autumn as much as any other time of year. The buff-red colors of the native grasses and the black seed heads of the meadow flowers jump out against the evergreen backdrops and blend seamlessly with the flaming colors of the trees. But when folks visit a wildflower meadow, they generally want to see wildflowers. In October, I worried, there would be very few of those.

As it happened, I was proven wrong. There was a profusion of flowers when the tour group visited, although not in the meadow. The forest edges that followed the long winding driveway to the clients' house were lined on that day with clouds of fluffy white blooms. Views into the sunlit gaps of the woodland interior were also bright with splashes of the same plant. The expert horticultural visitors were taken aback. What plant blooms this prolifically in October? How could I have planted it over such an expanse? How did I keep the planting so pure?

In fact, I hadn't done the planting. And if the tour group was amazed, it was in large part because they would never have considered encouraging this species in their own gardens. The star of this show was white snakeroot (*Ageratina altissima*), a common plant of the woodland edge, an opportunist that the fellows had surely come across before but had also as surely dismissed as a weed. A native weed, but a weed none-theless. Like any opportunist, the white snakeroot had planted itself.

It would not have been there in such numbers and purity, however, had we not taken certain actions during the two years that preceded the visit. Prior to bringing these woodland edge areas into our management regime they were covered with invasive brambles and vines. We eliminated this unwanted vegetation; not just the mature plants, but also the small seedlings that germinated once the competition from their parent plants were removed. In doing so, we had purged the seed bank (the repository of seeds lying dormant in the soil) of invasive species. As a result, the process of ecological succession was allowed to proceed as it would have historically, before the invasive species were introduced.

On this site, the species composition of the current stage of succession had been altered by the aggressive invasive plant species that had spread throughout it, crowding out or suppressing most of the natives. When we removed the invaders and unleashed succession, what had appeared in great abundance in the area vacated by the invasive species was white snakeroot in pure stands.

White snakeroot

Columbine

Canada mayflower

pages 51–55 Our garden star, white snakeroot, colonizes quickly after disturbance, then yields to longer lived, more aggressive species. Planting and management procedures guide the woodland into this next vegetative phase, which includes columbine, largeflower bellwort, golden ragwort, and other woodland herbs.

Golden ragwort behind phlox

As avid gardeners, these visitors must have seen white snakeroot before, even if it had not made much of an impression. Its showy, pure white flowers set atop a mound of clean dark green foliage should make it a welcome addition to any natural area or garden, but typically it falls short. Why?

The answer lies in the vegetative context in which this plant is usually seen the wild. Its natural habitat is partly sunny woodland edges and gaps, just where it occurred on the Connecticut property. Unfortunately for white snakeroot, these are the same tumultuous zones where, due to abundant sunlight and frequently lack of management, invasive species are most aggressive. In these scenarios, white snakeroot generally produces only small patches or scattered individual plants that are leggy, leaning, and barely able to poke through the more aggressive invasive species.

No living thing looks good, no matter how inherently attractive, when it is in the process of being battered by thugs. Here, however, our selective alteration of the competitive balance—the removal of invasive species—resulted in a pure stand of the white snakeroot as it would have formed historically. Restoring the successional process had turned a garden outcast into a landscape star.

When native plant communities and natural processes unfold as they have naturally evolved over hundreds or even thousands of years, people usually interpret the results as beautiful. This was certainly the

Rhododendrons behind amsonia

case with our visitors from the Garden Conservancy: a happy ending to our story.

But not so fast.

Had the visitors who were transfixed by the snowy sweeps of snakeroot stooped to look at the ground around their bases, they would have found clear hints of the future, one that would look very different. Below the white snakeroot they would have found small seedlings of slow-growing and more competitive plants that were preparing to form the next wave of vegetation. Goldenrods, in this instance, with scatterings of spotted geranium, white wood aster (*Eurybia divaricata*), and Christmas fern (*Polystichum acrostichoides*). While the Garden Conservancy fellows were admiring the status quo, I was already plotting how I would keep the ultra-vigorous (if native) goldenrods from overrunning the whole area and how I could promote the geraniums, asters, and ferns as the next stage unfolded.

What's the moral of this story? That a garden, at least an ecologically managed one, is as much a process as a place. Restoring and unleashing the natural dynamics can turn struggling natives into stars, but they won't dominate more than one act in an ongoing drama. If those Garden Conservancy fellows revisited that north-western Connecticut garden this autumn, they would find it transformed but equally lovely, I think, in its current incarnation.

Just as you cannot step into the same river twice, if you are gardening ecologically, you won't find the same garden on two different visits.

above White snakeroot, found in eastern and central North America, seeds itself around readily. This perennial is usually considered a weed, but it is quite beautiful when liberated from the smothering effects of invasive species.

Largeflower bellwort at center

THE GARDEN
ECOLOGIST'S PRIMER

The vocabulary of ecology plays a fundamental role in this design approach. In part, this is a matter of efficiency: ecological terms such niche, habitat, and ecoregion communicate a great deal of precise information very succinctly. But there's a far more important reason for developing a fluency in basic ecological terminology than just becoming comfortable with professional shorthand. In the process of learning these terms, you'll also learn a different way of looking at, understanding, and responding to the landscape.

When, for example, the area including your garden site becomes a habitat or even a series of related habitats, that means you have begun to understand that the conditions found there dictate what plants and animals will thrive in your garden. And when, instead of a collection, the plant inhabitants of your garden become a community, you are beginning to understand that the species that live in a habitat are not just accidental neighbors, but that they by definition tolerate the same conditions, and in some cases they may interact in a positive way.

As you make this sort of terminology a routine part of your design, planting, and management work, you will move beyond the traditional view of garden design as a solely aesthetic act. For you, this process will also become a program for the creation of an ecologically functional system. When you begin to think of your garden as an ecosystem, you have moved beyond the traditional concept of a garden as a collection of plants selected solely on the basis of flower color and foliage texture. You have recognized that your garden's vitality is based on the interactions of the plants with each other as well as the soil, topography, and local wildlife.

The time you invest in mastering a basic ecological fluency will not only help you to understand the chapters that follow, but allow you to see more deeply into and interact more effectively with the landscape where you live or work. Rather than arranging the ecological terms alphabetically, I group together related concepts and set them down in the sequence in which you will commonly encounter them during the design process. In this way, the information in each entry follows naturally from the preceding one, which makes the learning process simpler.

Waves of contrasting textures and hues define this wild garden.

ECOREGION

An ecoregion is the geographical area within which similarities of climate, soil, and types of landforms nurture particular associations or communities of plants. Determining the ecoregion of your landscape and including plants that are indigenous to that ecoregion is the essential first step to creating an ecologically based garden. Fortunately, the U.S. Environmental Protection Agency has produced detailed ecoregion maps of the United States and Canada, which can be found online (epa.gov/wed/pages/ecoregions.htm). Currently, the most generalized units (Level 1) define ten ecoregions in the continental United States; the finest-scale units (Level IV) include 967 ecoregion designations.

Garden centers and mail order nurseries typically offer a selection of plants based on the U.S. Department of Agriculture's Plant Hardiness Zone Map, which divides the United States into eleven distinct zones based on the lowest temperature experienced locally in an average winter. Using this as your only reference means that a gardener in Albany, New York (where the annual precipitation averages 39.35 inches), would cultivate exactly the same plants as one in Denver, Colorado (where the average annual precipitation totals 17.07 inches). Using your ecoregion as a guide instead ensures a far better adapted selection of plants.

Bunchberry dogwood (*Cornus canadensis*) provides an example of the need to consider the ecoregion where a species naturally occurs. This relative of the familiar native flowering dogwood (*Cornus florida*) tree is a spreading, low-growing shrub that forms a 6-inch-tall mat of dogwood-like foliage and, in late spring to early summer, white or greenish and sometimes pink dogwood-type flowers that are followed by bright red berries. It's an outstanding native groundcover, a plant anyone would want in the garden, and according to the USDA hardiness zones, it should grow in Philadelphia. In fact, it does not flourish there, because as a native of more northerly ecoregions, it cannot tolerate the heat of the summers. Nor does it thrive throughout most of the Rocky Mountain region, where summers may be appropriately cool but the summer rainfall doesn't meet the bunchberry dogwood's need for moisture.

NATIVE

One common error of native plant enthusiasts is that they tend to define native in terms of state or national boundaries; for example, describing a particular species of wildflower, grass, tree, or shrub as native to New Jersey or Virginia or Quebec. In fact, using politically drawn borders to describe where a plant is indigenous is commonly misleading.

Consider, for example, sundial lupine (*Lupinus perennis*), a prized perennial with lushly textured leaves and spikes of blue, sweet pea–like flowers. Although this species is listed as native to Delaware, in fact, it does not flourish in most areas of that state. Sundial lupine is a legume, a member of a group of plants that typically flourish on poor soils because their roots have the ability to convert atmospheric nitrogen into a natural fertilizer. As a result of this trait, you'll find sundial lupine on Delaware's coastal plain, where this plant can cope successfully with the infertility of the sandy soils prevalent in that region. In Delaware's Piedmont region, however, the soil tends to be a dense, impervious clay that limits the lupine roots' exposure to air, thus negating its peculiar advantage, and the species is rarely successful there.

In actuality, of course, plants don't pay much attention to state lines. The vegetation found in the coastal plain of New Jersey has far more in common with that of the Delaware coastal plain, and vice versa, than either coastal plain flora has in common with that of the state's Piedmont ecoregion. In this book I use ecoregion to define a plant's nativity. While it is by no means necessary to select only plants from within the ecoregion in which the garden is located, including at least a strong component of these highly adapted species will increase the garden's vitality and ultimately its self-sufficiency.

Bunchberry dogwood forms a ground layer under sheep laurel (*Kalmia angustifolia*) and northern bayberry along a roadside in Maine. At first glance, the three plants are so interwoven that they appear to be one.

HABITAT

For plants, a habitat is both a place and a set of environmental conditions. That is, it's the type of place that a given plant species naturally inhabits, such as a mountaintop or marsh. However, each habitat is also distinguished by a certain set of physical conditions, such as a particular soil type, the range of temperatures experienced in the place, the amount of moisture available to the plant, and the intensity of the sunlight. For example, a habitat may be the floor of open woodland with dry, acidic soils and sunlight penetrating intermittently through the day.

A landscape or garden may include several habitats: a corner shaded by trees and shrubs, for example, alongside an open, grassy area exposed throughout the day to full sunlight, perhaps backing onto a dry, rocky slope. Identifying the habitat or habitats of your garden, and using plants that naturally occur in that habitat, is the next step toward refining the selection of plants you will draw on.

An example of failing to consider habitat can be found in an experience I had early in my career with pink tickseed (*Coreopsis rosea*). When the plant was first introduced into the nursery trade some two decades ago, I was already making liberal use of a related plant, whorled tickseed (*Coreopsis verticillata*), a beautiful perennial with feathery foliage, a vigorous spreading character, and profuse yellow flowers. I particularly prized the horticultural cultivar 'Moonbeam' for its ability to bloom throughout the summer, an extremely rare characteristic for a long-lived perennial. That's why I greeted the appearance of pink

Pink tickseed grows in open sunny wetlands in select areas of the east coast, particularly in the coastal plains. This species will not thrive in garden situations if its specific habitat requirements are not met.

tickseed with so much enthusiasm. The plant looked identical to whorled tickseed in foliage texture, height, spread, and flower form. The only apparent difference was the color of the flowers. I could now create a lively mix of pink and yellow flowers atop a soft, uniform bed of foliage, and I quickly added pink tickseed to my plant palette. Unfortunately, if I planted ten, the next year I had seven, then three, then none. I quickly came to the conclusion that this was a bad plant and dropped it from my roster.

What I did not understand is that these two plants, while similar in appearance, came from entirely different habitats. My reliable yellow-flowered friend, whorled tickseed, commonly occurs in dry fields and open woodlands. My disappointing pink newcomer grows only on the moist fringes of open sunny wetlands. You can't have both conditions in one spot, so by definition one of them was going to die. Given that the habitat that supports whorled tickseed is much more common than that of pink tickseed, the pink was usually the one to drop out. My superficial observations regarding the similarities of these two plants failed to reveal the information I really needed to know to successfully grow pink tickseed. The questions I neglected to ask were "In what habitat does this plant naturally occur, and do I have a similar condition where I am working?" Pink tickseed was not a bad plant, I was a bad ecologist. And now, when the right conditions present themselves, pink tickseed appears again in my landscapes. But these days, it tends to stick around.

MICROHABITAT

Within any habitat, variations in topography, moisture levels, sunlight, or other features can significantly modify the environmental conditions. These are called microhabitats, and they often support a specific suite of plants that differ from those found elsewhere on the site. A north-facing slope, for example, is likely to be consistently cooler than adjacent south- or west-facing ones that receive more intense sunlight. Although all other conditions may be the same, different plants can often be found on each. A good example is Christmas fern, which tends to be much more prolific on north-facing woodland slopes than on south-facing ones, certainly in the mid-Atlantic states and southern New England. Once observed, this condition can help inform plant selection for the ground layer of a newly planted tree grove on a north-facing slope.

Microhabitats can also be the result of human activity, and such areas can present opportunities to grow plants that would not otherwise flourish in your garden. If I were determined to grow bunchberry dogwood, a northern species poorly adapted to the warmer climate of Pennsylvania, I would look for a stonewall that cuts across a north-facing slope. I would then set the bunchberry plant at the foot of its uphill side. Because cold air is heavier than warm air, it flows downhill like water and would tend to pool against the uphill side of the stonewall, creating a cooler zone, and because water seeps into the crevices between the wall's stones, more moisture would be present as well. This microhabitat may mimic the more northern ecoregion in which bunchberry dogwood is normally found and provide a rare opportunity to include this highly desirable plant.

PLANT COMMUNITY

A plant community is a group of plants that are adapted to the same habitat and commonly associate together within that habitat. Examples include dry oak–mixed hardwood forest in the Northeast and juniper–grass savanna in the Southwest. Using these plant groupings as a model for your plant selection not only ensures the plants' adaptability to the habitat in which they are being placed but also their adaptability to each other. After all, they have grown in close proximity for centuries, possibly resulting in a series of mutually beneficial interactions. In lay terms, they have learned to live cooperatively.

Modeling your plant selections on naturally occurring plant communities can have practical benefits beyond those derived from matching individual plants to their naturally occurring habitat. These can include reduced maintenance, particularly the reduction of weeding, which is usually the most time-consuming act a gardener performs.

Dense blazing star (*Liatris spicata*) grows naturally in wet meadows, and in this type of habitat it can put on a spectacular show, forming a cluster of stems and in mid to late summer raising its fluffy, cylindrical purple flower heads 2 to 4 feet high, or even more. A gardener seeking bold effects might decide to plant this species in a drift by itself. Even though it may thrive in its preferred wet meadow habitat, blazing star planted as an individual is likely to demand constant attention: its upright, spindly form makes it especially vulnerable to weeds. Mix it into a tapestry of native grasses and other meadow flowers, however, and its ability to force its way upward through low, creeping, groundcover plants and bushy lower plants enables it to add a strongly contrasting note of color and form, while the low-growing associates provide the dense weed barrier that blazing star can't provide on its own.

A plant community, however, doesn't always involve many different species. Woodland sunflower (*Helianthus divaricatus*) grows at the edge of moist woods, and this robust,

stoloniferous plant smothers adjacent plants, soon expanding into a large colony of just this one species. Essentially, this is a community of one. Unlike blazing star, which forms a dense cover only in concert with its plant associates, planting a single species drift of woodland sunflower in the garden can achieve a low-maintenance weed-suppressive composition.

Planting in communities differs from conventional gardening practice in a fundamental and very important way. Traditional gardeners commonly interplant species from many different geographical areas. What results is a menagerie of competing individuals that in essence just met the day you planted them. In contrast, if you base your design on a plant community, you plant combinations of species that have proven their ability to coexist in a balanced composition. That is, over time they adapted not only to the conditions of their habitat, but also to the presence of each other. This creates a true community, in which different species have adapted to occupy different niches so that they can coexist within the same area. This in turn means that the gardener isn't forced to intervene constantly, playing referee with battling species, as in a conventional garden.

Blazing star thrives in a wet depression in a designed meadow.

The tall spikes of blazing star fill in any available gaps between big bluestem and narrowleaf mountainmint (*Pycnanthemum tenuifolium*). In this community planting, every spatial niche is occupied, making the composition highly weed suppressive.

NICHE

In its original sense, a niche is a recess in a wall, in which a statue or some other object might be placed. An ecological niche is similar, in that it defines the spot and circumstances in which an organism, plant or animal, fits into a community.

An ecological niche can be spatial. A forest floor is a niche for ground-covers and ground-dwelling forest animals, just as the forest understory is a niche for shrubs and smaller shade-tolerant trees such as flowering dogwood, and the forest canopy is a niche for birds or tree-dwelling plants such as Spanish moss (*Tillandsia usneoides*) and lichens. Because all of these species are growing in the same area, one would think they are competing intensely. They may not be as much as you might think, because in many cases plants that occupy the same horizontal space are operating on different vertical layers.

An often overlooked type of spatial niche is the underground zone that a plant's roots inhabit, whether that is the surface soil through which Canadian anemone (*Anemone canadensis*) spreads its mat of rhizomatous roots, the mid-level that is inhabited by the deeper fibrous roots of little bluestem (*Schizachyrium scoparium*), or the deeper zone that is penetrated by taproots, such as that of butterflyweed (*Asclepias tuberosa*). Recognizing all of the spatial niches in your garden and finding desirable plants that will fill each one is essential to creating a landscape that resists invasion by weeds—in ecological terms, a relatively stable ecosystem. Leave a niche unoccupied and some opportunistic, aggressive plant species is sure to infiltrate it and, by overrunning or otherwise disturbing its neighbors, create opportunities for other invaders.

A community's different species also inhabit niches in time, and it is equally essential to recognize these and incorporate them into your planning process. The season that a plant is most actively growing represents one type of niche in time. In the midwestern prairie, bluejacket (*Tradescantia ohiensis*) grows in the spring and goes semi-dormant in the heat of summer. Indiangrass (*Sorghastrum nutans*) is a warm-season species that does not become active until the heat of summer. While these two species may occupy the same space, they are not in direct competition because their active growth periods occur during different seasons.

Niches in time can also be defined in years. Looking again at the midwestern prairie, blackeyed Susan (*Rudbeckia hirta*) is a biennial that may be a significant player in the first two years. Like all biennials, it forms leaves the first year, flowers and seeds the second year, and is gone by the third.

Short-lived perennials like lanceleaf tickseed (*Coreopsis lanceolata*) may replace the black-eyed Susan and dominate for a number of years until they are supplanted by slow-growing and long-lived species like white wild indigo (*Baptisia alba*), which may persist for a hundred years or more. All of these plants were likely present as seeds and seedlings during the early stages of the prairie's development. By growing at different rates and having differing life spans, they can avoid competing by occupying different niches in time.

Many popular, pre-packaged meadow seed mixes rely heavily on annual and biennial wildflowers. Such a mix typically produces a lot of colorful flowers the first year, but the subsequent niches in time are unfilled. These meadows inevitably degenerate into a weed patch due to the lack of late-stage competition. Excluding these short-lived species and relying only on long-lived ones may seem like a more prudent approach, but is also flawed. Long-lived species are generally slow to develop, and if only they are included in the early developing stages of the meadow, it will be open to weed incursion. A well-designed meadow seed mix includes long-, mid-, and short-term species, in effect filling all of the temporal niches and leaving little opportunity for weed incursion.

Weed suppression and the vitality gained from years of coevolved plant interactions are huge benefits gained from considering plant community and niche when selecting garden plants. But there are aesthetic benefits as well. Plants that have spatially and temporally sorted themselves through years of coevolution often display a balance, grace, and intricacy that are difficult to achieve in traditional gardens. After all, nature has been in the plant arranging business far longer than even the most accomplished garden designer.

NOVEL ECOSYSTEMS

Few ecosystems you will encounter in the more populated areas of North America have escaped transformation by human activities. In many cases, the ecosystem has been so changed by development for housing or commercial purposes, the introduction of invasive species, and changes to drainage and water sources that restoration of the ecosystem to something close to its historic state is impossible. Such a permanently altered, artificially created situation has come to be called a novel ecosystem.

In many instances, the garden ecologist has little choice but to work with the existing novel ecosystem. Doing so, however, raises some challenging issues. If an invasive species has become a part of the plant community there, should the gardener try to eliminate it? Eradication may be impossible, and the best that can be hoped for is beating it back and inserting a competitive native species into the resulting gaps.

An interesting development of recent years is the deliberate creation of planned novel ecosystems. James Hitchmough, a professor of horticultural ecology at Britain's University of Sheffield, has used his knowledge of natural ecosystem dynamics to create novel ecosystems with plants from many different regions for urban settings such as reclaimed industrial areas. His rationale for this practice is that he can create unnaturally intense displays of flower color and foliage textures that benefit city residents more than less spectacular ecosystems native to the region.

A front lawn becomes a meadow.

In its second year of growth, this designed meadow features a variety of short-lived species, including blackeyed Susan and foxtail barley.

In later years, little bluestem, coneflower species, and narrowleaf mountainmint are visually dominant. These long-lived species will persist indefinitely, as they have historically in undisturbed areas of the midwestern prairie.

GENERALIST VERSUS SPECIALIST SPECIES

Some plants flourish only in a very specific set of conditions or habitats. In ecological terms these species are called specialists, whereas those that grow in a wide range of ecological conditions are called generalists. Exemplifying the specialists is the pink lady's slipper (*Cypripedium acaule*). This hardy, native, ground-dwelling orchid flourishes only in semi-open woodlands in deep humus and acidic but well-drained soil under birch (*Betula* species) and other deciduous trees, and even there it requires the presence in the soil of a particular fungus to survive. This plant's exotic, large, slipper-shaped pink flowers attract many gardeners, but few gardens have the highly specific ecological conditions the pink lady's slipper requires, and efforts to impose the plant where these conditions don't exist invariably prove unsuccessful.

Most of the plants found growing in a vacant city lot fit into the generalist category, with tree of heaven (*Ailanthus altissima*), the tree that can grow from any crack in the pavement, being the supreme example of this. A more desirable, native generalist is the red maple (*Acer rubrum*), which flourishes in wetlands as well as poor, dry soils and almost everything in between. White wood aster is another example of this group. It grows in most woodlands in the temperate United States and Canada and consequently is easy to cultivate successfully. Plants in this category often provide the meat and potatoes of natural design. Highly specialized species, often the most prized species for connoisseurs of native plants, are used only when the appropriate habitat is present.

A switchgrass cultivar derived from an eastern ecotype stands rigidly upright in the Pennsylvania garden all season long.

ECOTYPES

When a species of plant is exposed to different conditions, over time it may adapt to each condition, so that populations of the same plant may differ markedly in appearance and preference for growing conditions among areas. Such variants within the species are called ecotypes.

Ideally, you should match the ecotype to your conditions. I was reminded of the importance of this when I planted the switchgrass (*Panicum virgatum*) cultivar 'Dallas Blues' in a Pennsylvania garden. 'Dallas Blues' can be a handsome plant, with blue-green foliage, extra-large flower panicles, and good winter color. It's also supposed to be upright, but in Pennsylvania it literally flopped. My suspicion is that this plant is an ecotype of switchgrass suited to the dry climate of the American Southwest, and that the moister climate of Pennsylvania provoked it into unusually tall growth that it could not support. Had I planted an ecotype that originated in some area with conditions more like those in Pennsylvania, the plant would have been better adapted to the rainfall levels to which it was exposed and remained standing.

Ecotype variations occur not only in different regions but also in different habitat types within regions. This can be illustrated by an experience that was related to me by Jim Plyler of Natural Landscapes Nursery in West Grove, Pennsylvania. Jim was collecting seed from wild populations of Catawba rhododendron (*Rhododendron catawbiense*) in the Smoky Mountains, and he found that as he ascended to the top of the higher peaks, the species changed appearance, with the shrubs exhibiting a scraggly branching structure with smaller, sparser foliage than those on lower, more protected slopes. He assumed that the difference in appearance was simply the effect on the rhododendron of the harsher conditions at high elevations. Yet when he returned home and grew plants from

the high-elevation seeds, he found that the mountain bald off-spring had the same scraggly, sparse look as the parents, while offspring from the low-elevation plants remained lush and full. He had collected seed from two distinct ecotypes, even though they were growing on the same mountain.

How does a gardener who wants to purchase plants with optimal adaptability to the site get his or her head around that? While very few nursery professionals currently make an attempt to stock local ecotypes or note the region from which their stock of a particular plant originated, specialized native plant nurseries are beginning to do so. This trend is likely to increase, and when it does, finding a local ecotype will be as simple as running to the corner native plant nursery. Maybe someday even microhabitat-adapted ecotypes will be available, such as that which Jim Plyler found on the mountain bald.

COMPETITION

When two organisms in close proximity both target the same resources of water, soil nutrients, and sunlight, they are said to be competing. Commonly a more aggressive and robust or more persistent species will outcompete the other, reducing its numbers within the plant community or eliminating it altogether.

When plants knit tightly together, as in this garden, there is little exposed soil in which weeds can become established.

In so many gardens, competition is eliminated as each plant is given its own space with open mulched area in between. This wide open space is nice for the plant as it doesn't need to compete with its neighbor, but it is also nice for the weeds, which take advantage of the lack of competition to aggressively establish in these open areas. In a community-based garden, dense and intermingled vegetation forms a solid ground layer, inhibiting most of these weeds. The planted species, however, are now operating in a competitive environment, meaning they need to expend more energy to survive. Consequently, a plant that may persist outside of its normal habitat in a relatively noncompetitive traditional garden environment may quickly drop out in the more competitive natural garden. This means that the garden ecologist has an increased need to match the plant with the proper ecological condition to obtain the dual goals of weed suppression and successful plant establishment.

SUCCESSION

Thus far, I've described where plants occur naturally (ecoregion, habitat) and the patterns with which they arrange themselves (plant community, niche). In my discussion of succession, I take this conversation to a whole new level, one that considers processes of compositional change over time, something that occurs in almost all natural landscapes. Visit an unmanaged open field today and you may see grasses and flowers dominating the mix. Return in ten years and some of the herbs have dropped out, supplanted by shrubs and young pioneer trees. Ten years later you may find the field herbs largely gone due to reduced light conditions from the now dense tree canopy, with shade-tolerant ferns and woodland wildflowers beginning to replace them. While this is by no means the only progression that can take place, succession is generally defined as the process by which a biological community—all the plants, animals, and other organisms—tends to change composition over time, usually in a somewhat predictable fashion.

Depending on where you garden, ecological succession may naturally progress in different directions. East of the Mississippi River, for example, and in Canada's eastern provinces succession on most sites tends toward an eventual condition of woodland. In the southwestern prairie states, succession is more likely to progress toward grasslands, whereas in parts of California shrubland may be a common successional direction.

Most traditional gardeners fight this process and do all they can to keep the original planting static. A garden ecologist looks for opportunities to use succession in the landscape—ride the wave, so to speak, or at least guide the wave—instead of constantly battling to hold it back.

One example is the use of a meadow planting to foster the development of a woodland landscape. Newly planted trees take years to form a canopy capable of providing sufficient shade for woodland understory plants. Densely seeding meadow plants around the young trees can be a cost-effective and aesthetically satisfying way to stabilize the site until the trees mature. It certainly beats large areas of bark mulch or frequently mowing and weed whacking around each tree. As the trees grow and shade the site the meadow will thin out, but you can replace it with ferns and woodland wildflowers or let this process occur naturally. In the meantime some of the original sunny meadow species will find refuge in woodland openings and sunnier edges as the trees mature and the shade beneath them deepens. You have essentially mimicked the natural process of succession; and instead of mulching or mowing while your woodland developed, the birds, the insects, and you have benefited from a diverse and beautiful meadow.

DISTURBANCE

Ecologists categorize anything that causes a temporary change of the average environmental conditions in an ecosystem as a disturbance. This can be a natural event such as a fire, flood, or storm, but it also includes human-induced events such as trampling or logging. On the regional scale, the introduction of invasive plants and animals and climate change can also be considered forms of human-induced disturbance.

A major disturbance commonly has the effect of resetting the successional clock, by eliminating or damaging the existing vegetation and providing opportunities for earlier pioneer species to re-emerge. The process then marches forward, either retracing its steps or moving in a different direction if conditions have changed significantly.

There are two common types of disturbance: vegetative and soil disturbance. Vegetative disturbance only affects the aboveground portion of the plants that are present. Examples include mowing or grazing and fire. This type of disturbance will often alter the relative abundance of plants that are already growing on the site. It may cause the re-emergence of species that were dominant in earlier stages of landscape's development but are now on the wane due to competition from stronger late-stage species. Soil disturbance is a deeper form that not only affects the aboveground vegetation, but plant roots and soil as well. Trees that are uprooted from a storm or the ravages of a bulldozer are two examples. This type of disturbance fosters germination of seeds because of the removal of competitive cover and the creation of a loosened seedbed. The plants that emerge from this type of disturbance are often species of an even earlier stage that have long faded from the scene, but can now re-emerge from seeds deposited by the original plants.

Gardeners are constantly creating disturbances, even if they don't think of it in those terms. An example is the simple act of weeding. A good gardener carefully pulls a weed to ensure that all of the roots have been removed. This procedure creates a soil disturbance, which as we have seen prompts the germination of more weed seeds. The net result is a never-ending cycle of weeding. There is a way to break this cycle, but it involves controlling the weeds while avoiding disturbance.

r- AND K-SELECTED SPECIES

This may be the ultimate example of off-putting professional jargon. What could sound drier than "r- and K-selected species"? But this is an instance where jargon is useful in communicating much important information succinctly.

A plant's reaction to disturbance—some find it devastating, whereas others find in it an opportunity—distinguishes which of these two categories the species belongs to. Understanding and exploiting the difference between these two classes enables gardeners to protect their landscapes against weed invasion and create quick, spectacular displays, as well as long-lived, more stable plant communities.

r-selected plants are species that have evolved to grow successfully in habitats that recently experienced some major disturbance, such as birch. These plants, often called pioneer species, are genetically programed to flourish in precisely such circumstances. They are fast growing and reproduce quickly, so that within a matter of months or even weeks they can flood the stripped habitat with offspring. Often, they are annuals, plants that flower and set seed only once before they die, although r-selected plants also include many short-lived perennials, such as white snakeroot. Establishing quickly and having a short life span is key to their success; r-selected plants don't invest their

The white spires of Culver's root and red plumes of queen of the prairie (*Filipendula rubra* 'Venusta'), two examples of K-selected species, thread through a mosaic of other wildflowers and grasses in a mature seeded meadow.

energy in long-term growth. After all, in the disturbance-prone habitats in which r-selected plants evolved, life expectancy is limited by the frequent recurrence of disaster. Because they emphasize reproduction rather than longevity, r-selected plants tend to be shallow rooted, but their bloom is often extraordinary. They're similar to a person who lives high off the hog on credit cards but crashes and burns when the bills come due.

K-selected plants, like blue wild indigo, are slower growing but persistent and put much of their initial energy into root growth before they produce flowers. These plants commonly produce fewer and larger seeds, such as the acorns of oak. Unlike the r-selected plants, they put their money in the bank and wait until they have adequate resources before they start living large. An example of this type of plant is the native prairie perennial Culver's root (*Veronicastrum virginicum*), which may take ten years before displaying a visible presence in a seeded meadow. But once it is there, it is there to stay.

These plants tend to inhabit more stable habitats such as mature woodlands or established meadows, where major disturbances are uncommon. In a setting of this kind, K-selected plants can afford to invest in slower, more expansive growth. K-selected plants flower and set seed less prolifically than their r-selected counterparts, but typically rebloom year after year. Because they live in more stable habitats where the plant community is richer and more diverse, K-selected plants are more likely to form mutually beneficial associations with their natural neighbors. As a result, when planting K-selected species, it's important to plant their natural companions as well. Treat them as just one element of a complex.

These different patterns of growth dictate different types of usage by the designer. Planting K-selected plants is a long-term investment. They should not be expected to spread rapidly, but on the other hand, if the species you plant are adapted to the conditions on the site, they will contribute an enduring presence and display. In contrast, r-selected species make a very useful patch. Inserting them where a landscape has suffered disturbance is an easy way to fill a vacant area quickly that would otherwise invite colonization by invasive species. Pioneer species are also a means to nearly instant gratification in the early stages of implementing a design. As the visitors to that northwestern Connecticut estate discovered, r-selected plants such as the white snakeroot can mount a breath-taking display. And as the landscape matures, instead of persisting as weeds, they disappear almost as quickly as they spread, giving way naturally to their K-selected successors and surviving only where human or natural disturbance creates a gap in the living fabric of the garden.

COLONIZATION

Ecologists do not regard a plant as being successful in the landscape simply because it survived. It is only considered successful if it reproduces in that landscape. Consequently, population increase—the multiplication of a species within a habitat—is the real test of a plant's adaptation to the conditions there and the truest form of longevity.

A plant may proliferate by seed or vegetatively by spreading roots and stems to colonize a new area. Plants tend to put their energy into one form or another. Pennsylvania sedge, which spreads by rhizomatous roots, produces little viable seed, whereas those sedges that spread by seeds often remain as an individual clump.

Understanding how a particular species proliferates is essential to using it appropriately as an element in a garden design. For example, plants that reproduce vegetatively may do so by extending their roots or underground stems (called rhizomes) outward to enlarge the original clump. They tend to densely occupy space above- and belowground and consequently can be quite weed suppressive when planted as a single species. Hardy ageratum (*Conoclinium coelestinum*) is an example of this type of species. The soft blue flowers it bears from late summer until the first hard frost are a very visible attraction, but it's also a useful groundcover, as it rapidly forms a dense monoculture over a large area.

Some other plants proliferate vegetatively by stolons, aboveground shoots. Typically, a stolon sprouts from one of a plant's existing stems and then threads its way through or over neighbors until the stolon's tip reaches a gap in the surrounding vegetation, whereupon it roots into the soil to start a new plant. Stoloniferous plants are particularly useful as groundcovers in a mixed planting, as they tend to coexist without harming neighbors while methodically plugging gaps in the surrounding vegetation that are otherwise likely to sprout weeds.

A crucial factor in creating a self-perpetuating garden is using plants that disperse their seed effectively. Plants accomplish this by various means, including wind, gravity, birds, mammals, water, and even ants. Knowing how a species' seeds are distributed about the landscape helps to govern its role in the garden.

The seeds of early meadow-rue (*Thalictrum dioicum*), for example, are encased in a silver fluffy cluster that can drift along in the wind for long distances. Such a plant needs little help in colonizing, as the installation of just a few plants soon produces progeny throughout. In contrast, species whose seeds are dispersed by gravity, simply falling to the ground around the parent, as is the case with the spotted geranium, may need assistance to colonize effectively. To ensure a broad distribution of this plant, you either need to plant many individual specimens around the landscape—although you can also wait for a single well-sited specimen to surround itself with seedlings and then transplant these offspring—or you can harvest seed and sow it in appropriate spots. Starting with purchased seed, it's worth noting, rarely works well with spotted geranium because, like that of most woodland wildflowers, the life of the seed is very short. By the time it is collected and stored by the supplier, shipped to you, and eventually planted, it will most likely no longer be viable. It's better to plant your own seed source and have seeds hit the ground the moment they are ripe, as occurs in a natural setting.

Statuesque cup plant (*Silphium perfoliatum*) seeded itself into this garden from plants growing in a nearby meadow. Through the serendipity of nature, it now serves as a strong focal point in the garden.

Birds may also assist in distributing seeds throughout your garden. Of course, given that these creatures are so highly mobile, there's a real chance that the fruits they eat from your southern arrowwood (*Viburnum dentatum*) could be deposited later in the day on your neighbor's property. To prevent this, keep the birds on your lot as long as possible by providing for a fuller range of their needs: don't just attract the birds by planting berry-bearing plants, furnish a water source and cover to protect the birds from predators. That way, when "nature calls" the bird is more likely to answer on your property, instead of eating and running.

Some wildflower seeds, such as those of bloodroot (*Sanguinaria canadensis*) and trilliums (*Trillium* species), have oil-rich elaiosomes, fleshy appendages that ants prize as food. Worker ants harvest the seeds and carry them back to the nest, which effectively plants the seeds in nutrient-rich environment. These are specialist plants that won't survive in just any soil condition. If you are lucky enough to have such a condition, don't mess it up by double-digging the planting area, adding fertilizer, or using insecticides that kill the very insects that are needed to distribute the plants.

Planting a diversity of plant species, in nature or a garden, is desirable. If one species fails, another is there to take its place. Similarly, using plants with a diversity of colonization strategies increases the chances that when a gap appears, regardless of its position in the landscape or its ecological condition, something desirable will fill that space. And what self-respecting gardener would turn his back on free plants?

SENESCENCE

Senescence is the natural aging process that occurs in both individual plants and in plant communities and ecosystems as a whole. This means, of course, that plants will die, some sooner and some later. Traditionally, gardeners have regarded senescence as an enemy, going to great lengths to extend the lives of individual specimens and populations even when the plants have proven poorly adapted to the site, have been overtaken by succession, or are just at the end of their natural life span. Though no one cares to lose a stately and ancient oak tree, an ecological garden can include the kind of change that senescence brings and the evolution it fosters within the landscape.

The indigenous landscape is a constantly changing system composed of plants, animals, microorganisms, and soils. Plants are not isolated entities, but participants in a system constantly in flux. The stage of a meadow when populated mostly by annuals may last for only one year following the disturbance, whether natural or man-made, that presented these grassland plants with the opportunity to grow. If succession isn't derailed by management, the stage when perennials dominate a meadow may last for ten years before yielding to pioneer forest species. Should we lament the loss of the meadow annuals? No. The garden ecologist recognizes that their time has passed and that each successional stage that follows will bring a new and exciting experience.

Planning for senescence can take many forms, depending on the type of habitat and ecosystem and the degree of control you wish to maintain. When planting a woodland, for example, you could plant not only fast-growing, pioneer species such as birches, but also intermingle with these more enduring species such as northern red oak (*Quercus rubra*), so that when the short-lived birches die, the oaks will be there and ready to fill the resulting voids. When creating a meadow, garden ecologists plant fast-maturing, fast-senescing annuals and biennials to provide quick cover and lock out weeds. At the same time, they plant more enduring perennials that will take over as the annuals and biennials senesce and disappear.

Even dead trees continue to contribute to the surrounding ecosystem. Leaving a standing trunk, known as a snag, can serve as a magnet for insects and the birds that feed on them for many years. At times I have even highlighted particularly sculptural snags and made them an aesthetic feature of the landscape, or planted new trees around them to symbolize the inevitable rejuvenation that senescence will ultimately bring.

INITIAL FLORISTIC COMPOSITION

When the field of ecology was first being developed as a science, ecologists believed that disturbance left a site as something similar to an ecological blank slate, and that the ensuing vegetative changes were largely determined by those seeds that were deposited from the surrounding area prior to each stage of ecological succession.

In an earlier chapter I profiled the work of the ecologist Frank Egler. It was typical of Egler's genius—and suggestive of why he could not keep an academic position within a university—that he was not afraid to contradict the prevailing view. Based on observations made over decades at his own property, Aton Forest, Egler developed a competing theory called initial floristic composition. His theory stated that most of the plants that grow on a site in the years following disturbance or release from cultivation were already present on the site, either as seeds, seedlings, or small plants. Unlike traditional succession theory, he believed that the seeds deposited after an initial vegetative cover was established were unlikely to germinate in any profusion due to the competition from existing vegetation.

This may seem like a minor distinction, but it has huge implications for garden establishment and ecological restoration. In the early stages of a landscape's development, which usually includes some form of disturbance, the plants that proliferate as a result of that disturbance are often a mix of desirable and undesirable species. If the undesirable plants are quickly removed before they can become pervasive, the desirable ones will likely dominate. Because newly deposited seeds are not likely to germinate with any profusion, once the initial work has been completed, managing the landscape over time should be relatively easy. In effect, the process of plant editing that was performed at the outset put that landscape on a vegetative trajectory that favored the desired species and was highly resistant to invading weed species.

Egler substantiated this theory by conducting ecological research at Aton Forest, his property in Norfolk, Connecticut. Woodchuck Hill is a 7-acre field there that had been mowed for hay. After closely observing the vegetation in the field, Egler noticed that beneath the dominant grasses were numerous but small individuals of low shrubs including

In this woodland garden, two large dead trees were deliberately left, one standing and one prone, to provide habitat for birds and invertebrates while giving a visual reminder of the site's prior vegetation and nature's pervading cycles of death and renewal.

lowbush blueberry, native steeplebush and white meadowsweet (*Spiraea tomentosa* and *Spiraea alba* var. *latifolia*), and common juniper (*Juniperus communis*). Because these native species were already present naturally on the site, albeit mainly in the understory below the grass, he decided to manage for them.

His first step was to cease mowing, a form of disturbance. As I discussed earlier, disturbance tends to revert or preserve vegetation at an early stage of succession. Because shrubs can eventually outcompete herbaceous species, ceasing to mow, Egler reasoned, should allow the shift to shrub species to occur, and it did. But while the shrubs overtook the grass, trees were also beginning to emerge in profusion. The field was succeeding to forest. Because preserving the area as an open space was one of Egler's goals, he began removing trees, one by one.

This was a substantial undertaking, even though the trees were still fairly small. However, once he completed the initial work, the number of trees that continued to invade the field dropped precipitously, and the native shrubs became the overwhelmingly dominant cover. According to his theory of initial floristic composition, the majority of plants that appear in an ecosystem over time were there from the onset of disturbance or release from cultivation, and new seeds entering the system after cover has been established have a very low rate of survival. Once Egler set the vegetative trajectory toward shrubland by ceasing to mow and removing the initial flush of trees, it became a relatively stable and easily managed landscape. Egler managed the field by simply removing the occasional tree for more than fifty years, until he passed away in 1996.

To understand the magnitude of this achievement, one need only look at the strip of 60-foot-tall woodland that surrounds Woodchuck Hill. Back in the 1940s, this strip was also part of the hayfield; Egler purposely left it unmanaged to serve as the control element in his management experiment. Egler didn't have a crew to manage this 7-acre expanse; he did all the work himself (today his colleague John Anderson continues this work). Using brains rather than brawn is a central theme of this book, and Frank Egler's work at Aton Forest and on Woodchuck Hill is one of best examples of that maxim that I have come across.

This Allentown, Pennsylvania, estate provided my first opportunity to work on a truly large scale. The plan included more than 3 acres of native gardens and 30 acres of seeded meadow. A pool was designed as a path leading from the gardens to the wild landscape. The scale of the project demanded that ecological patterns and processes be integrated into every aspect of the design, implementation, and management if a low-maintenance landscape were to be achieved.

DESIGN

THE PLANT THAT WANTS TO BE HERE

It now seems ironic that in 1831, when the Massachusetts Horticultural Society established Mount Auburn Cemetery, Bostonians would provide a more beautiful recreational ground for the dead than for the living. In fact, Americans of that era wouldn't spend tax dollars on public parks. Huge battles had to be fought to obtain funding for such projects as New York's Central Park or Boston's Public Garden, and these weren't won until a generation later. But a cemetery such as Mount Auburn, which extends over 175 acres on the border of Cambridge and Watertown, just across the Charles River from Boston, could fund its own construction by the sale of burial plots. Designed as a garden cemetery, the first of its kind in the United States, Mount Auburn was lavishly adorned with ornamental plants. Today it functions as an impressive, if unofficial, arboretum with its collection of some 16,000 plants of 1700 different types. It is also an island of nature within the tenth most populous metropolitan area in the United States.

I had come to Mount Auburn to participate in a charrette, a three-day design brainstorming session in which it was hoped the cemetery could find directions to move away from its dependence on turf and toward a more environmentally sustainable groundcover. What I found as I toured the grounds was that even after more than 180 years of landscaping with imported plants and the land's previous use for farming, remnants of native flora remained.

Observing what vegetation occurs naturally on a property, whether in unmanaged areas or planted garden displays, is always an integral part of initial site analysis. If a naturally occurring, desirable species establishes on a site without any human assistance, it is obviously well adapted and should be strongly considered as a component of the design. These species not only advocate for themselves, but also for the other plants with which they commonly associate, even if these associates are not present at the time of the site analysis. If conditions suit one member of a plant community, they will likely suit other members.

In this case, the plant that caught my eye was Pennsylvania sedge (*Carex pensylvanica*). I observed the plant growing unplanted along the wooded slopes of the cemetery's Violet Path, in the shady ravines of Consecration Dell, and as we circled the meadows at the base of Washington Tower. This meadow had been planted with native grasses and perennials. Pennsylvania sedge had not been one of them, and yet I found it well established there below a maple tree. In fact, this sedge was the most prolific component of the composition, knitting together the planted species that were present but had not formed a solid cover. That finishing touch had been achieved thanks to the unplanted sedge, and the foresight of the Mount Auburn staff, who didn't treat this volunteer as a weed and remove it, which would have been its fate in most traditional gardens. Clearly the sedge had found a niche.

This is not surprising, as the type of habitat in which this plant is most commonly found in the wild is similar to that found at Mount Auburn: open shade (which the cemetery offers because of the tree planting), acidic soil,

Eastern daisy fleabane, a native annual widespread throughout North America, has seeded itself into turf at Mount Auburn Cemetery. Management personnel have mowed around it, allowing this pop-up meadow to provide temporary interest.

and well-drained conditions thanks both to a relatively porous soil and to a hilly topography. Confirmation that conditions were favorable to the sedge could also be found in the fate of two other species that had been planted in the cemetery.

One of these was mountain andromeda (*Pieris floribunda*), which although it doesn't grow naturally farther north than Virginia, had been planted and was thriving in many areas of the cemetery. This shrub demands the same kind of well-drained, acidic soil and open shade favored by Pennsylvania sedge. In my experience with this shrub, I had found it to be finicky and unreliable. In fact, it was just a specialist that thrived only in very specific conditions. These conditions were present at Mount Auburn, and there the plant was anything but finicky.

The other confirming plant was one notable for its lack of success. This was maidenhair fern (*Adiantum pedatum*), which naturally occurs on limestone-derived (calcareous) soils with an alkaline pH. With its delicate foliage texture and the circular arrangement of its fronds, maidenhair fern is a popular garden plant and it too had been planted extensively in the cemetery. Its meager appearance, however, suggested that the pH of the soil was in the acidic range and better suited to the sedge.

Although I wouldn't have encouraged the cemetery to limit its planting in converting turf areas to a monoculture of Pennsylvania sedge, I did recommend it in my subsequent report as a potential backbone of a lawn alternative, one whose diversity could be enhanced by interplanting with low-growing flowers as well as other sedges (*Carex* species) with similar aesthetic characteristics and habitat preferences. As it had already done under the maple by the meadow, the Pennsylvania sedge could bind together the other players, serve as the primary weed suppressor, and create a lush natural texture for the ground plane.

Another advantage of the sedge as a lawn alternative, I suggested, would become apparent in post-planting management. When conducting site

Cascading Pennsylvania sedge with intermingled ferns and wildflowers provides a low-growing lawn alternative at this residential property.

analysis, it is just as important to inventory undesirable species as desirable ones. You have to know what your enemies are going to be. Virtually every urban and suburban space in North America has been colonized by invasive species, and Mount Auburn was no exception. Black swallow-wort (*Vincetoxicum nigrum*), Japanese honeysuckle, and Tatarian honeysuckle (*Lonicera tatarica*) were among the invasive species I observed there. Although I don't generally endorse the use of herbicides, they can be invaluable for eliminating invasive plants, especially if a lawn conversion will create planted areas that are too big to hand weed. Sedges aren't affected by the herbicides that specifically target grasses and broadleaf plants and so are immune to the materials

that would be used to address invasive species. This represents labor savings as well as the ability to control the composition, particularly during the early, more volatile establishment period.

This needn't become a permanent chemical dependency, however. Pennsylvania sedge is a very competitive groundcover due in part to its aggressive, rhizomatous root system. Once densely established, it is generally successful in suppressing further weed seed germination, so that the use of herbicides could be sharply reduced or even eliminated over time. If other species of *Carex* besides the Pennsylvania sedge were included in the planting, the resulting diversity would provide insurance against future insect or disease

Through its rhizomatous growth, Pennsylvania sedge colonizes the ground layer in this deciduous woodland, weaving amid ferns and tree saplings and providing a base matrix that has both beauty and ecological function.

outbreaks, without lessening the effectiveness of further herbicide applications if they were needed.

By day three of the charrette, I recognized what might prove another virtue of the Pennsylvania sedge in this particular setting. I asked the cemetery personnel about what kinds of activities had occurred in the past in various areas. In most places where the staff had disturbed the cover and soil, as well as in areas where the vegetation had been left unmanaged, Pennsylvania sedge had colonized and increased its numbers spontaneously. Would this also be the case in areas slated for conversion from turf? Would managed neglect favor the sedge over the turf grass and broadleaf weeds? I suggested experimenting with several measures, including raising the mowers to above the sedge height of 15 to 18 inches. This would cut any turf grasses or broadleaf weeds that grow through the sedge, but never the sedge itself. Over time this would favor the sedge, because it would not have to waste energy resprouting. The temporary use of selective herbicides could also be

part of the program to discourage plants other than the sedge during the establishment phases. Finally, because the sedge is adapted to a more acidic soil pH than turf grasses, the soil could be acidified through applications of sulfur. I also recommended proceeding with caution and testing all of these measures on a small scale before applying them more generally.

Over the three days that I observed the ecology of Mount Auburn, my thinking on their lawn alternative approach had evolved from planting Pennsylvania sedge to managing for the natural recruitment of Pennsylvania sedge. At this point, I don't know if the cemetery management will adopt my recommendations. What I do know is that they were not really my invention, but just my careful reading of what was already there. And that is the true secret of effective site analysis.

SITE ANALYSIS

WHERE ARE YOU, ECOLOGICALLY SPEAKING?

It's the most natural thing in the world, yet it's a fatal mistake. So often, when gardeners begin planning a new landscape or even an extension of an established one, they start the project with a detailed mental image of the result they want: some memory of an Eden that they carry forward from their childhood, perhaps, a garden they saw in a magazine, or simply a collection of favored plants. Then, without first investigating what their landscape's natural potential and proclivities may be, the gardener tries to impose this vision.

I've done this myself. But this sort of beginning is like starting a first date with a monologue rather than a conversation. Just telling someone all about yourself is not, typically, an effective way to initiate a meaningful, mutually rewarding relationship. Nor is it a good way to begin your relationship with a landscape.

Instead, begin with a question: Where are we, ecologically speaking, in both space and time? In other words, what are the existing environmental conditions of the property? What plants naturally inhabit those conditions? And equally important, but rarely considered, what natural processes of change will affect the landscape efforts going forward?

But how does this sort of site analysis differ from one performed by traditional designers, which would commonly examine soil type, light exposure, and moisture levels? Although these characteristics are certainly relevant, the relationships between plants and their historic environments run much deeper.

Plants have evolved adaptations for survival in the environments in which they live for hundreds or even thousands of years. Lady's slippers (*Cypripedium* species), found in woodlands through most of the United States, have evolved with specific soil fungi and will not persist without their presence. Indian paintbrush (*Castilleja* species), the spectacular orange wildflowers that blanket the shortgrass prairies of the southwestern United States in late winter, are associated with the neutral or high pH soils commonly found there. These plants are also found in the eastern United States but are not as ubiquitous, growing only in isolated pockets where soils derived from limestone exhibit higher pH than is typical in the east. Consequently, a gardener in the Texas hill country can plant Indian paintbrush in

This garden sits on a vein of limestone bedrock that yields soils with high pH. The design thus specified plants that occur naturally in more alkaline soil conditions.

a wide variety of locations, whereas a gardener in Connecticut is not likely to succeed with these plants unless the property happens to have an atypical pocket of limey soil.

Or consider meadow beauty (*Rhexia virginica*), an attractive perennial with profuse pink flowers that bloom for a long period in the summer. The basic cultural requirements of this plant are sun and lots of moisture: basically a wet meadow plant. Given the epithet of its species name (*virginica*), it would seem like a can't-miss selection for a sunny wet garden in western Virginia. But meadow beauty inhabits the wet meadows of the coastal plain in the eastern part of the state. These environments have well-drained sandy soils, with their wetness derived from a high water table. A wet meadow in the Piedmont regions of western Virginia, the home of our gardener, likely has a dense clay soil, its wetness the result of a low-lying position in the landscape and the poor drainage capacity of the clay. Both are wet and sunny, but only one will support meadow beauty. Clearly in this and many other cases, understanding where you are, ecologically speaking, and using that knowledge to place plants in the appropriate ecological context will significantly increase the likelihood that the plants you introduce will succeed. And because plants in the wild often compete for the same space, the more your plant introductions succeed, the less opportunity there will be for weeds to do the same.

All of these examples are spatial in character and illustrate the importance of matching the plant to the habitat in which it grows naturally. In nature, however, plant compositions almost always change over time. In some cases these changes unfold gradually as a result of normal natural process. In other cases change comes abruptly as a result of a natural event, such as a tree blowing down in the woods, or a man-made one, such as a bulldozer clearing for a development. How this change has unfolded on a site has a strong influence not only on the current plant composition, but on the direction that the plant composition will tend toward in the future. In addition to spatial analysis, temporal analysis is also needed to maximize your ability to influence that future change. When landscapes are designed as static plant compositions, this information may not be all that critical. But if a landscape is conceived as a changing composition that piggybacks upon natural processes, it is crucial.

Consequently, there are three key questions to ask:

- What plant composition likely preceded the current vegetative state?

- What human or natural occurrences affected that composition?

- What changes would likely occur if all maintenance activity ceased?

Because human disturbance is so widespread today, retrospective analysis is very important. Are there any remnants or other clues that suggest what the previous incarnation of this land was, before it was bulldozed or otherwise transformed for its present use? Are the trees covering the site representatives of a mature woodland, or are they pioneers that sprang up a decade ago when a farmer removed grazing animals or retired his plow? Are the grasses remnants of native grassland, or are they relics of the site's use

as a hayfield, or even the remains of an abandoned lawn? Knowing accurately where the site has been and where it is now will suggest where it will tend to go. And that is the basis for creating a design that works both ecologically and in human terms. What's more, the results you achieve in this fashion will be something intrinsic to the site.

By focusing on their own vision, traditional designers miss the subtler clues that are crucial to developing a landscape that will function well ecologically. It is the landscape itself that tells us what will thrive there and which plants, once introduced to the site, will propagate themselves to form a self-sustaining flora.

The landscape plan for this southeastern Pennsylvania property included the planting of meadows, shrublands, woodland groves, and woodland edge. A reliance on careful site analysis optimized the chance that plants would be placed where they are most likely to survive and be easily managed.

SITE ANALYSIS

While working your way through this site analysis, keep in mind that none of the answers you arrive at initially should be regarded as final. That's because the process of getting to know the site progresses simultaneously with the design process. Your understanding of the existing landscape will inevitably evolve as you work on it, as the landscape's responses to your inputs will deepen your knowledge of the place with which you are working. Likewise, your design should also evolve as you get to know the site more intimately.

DETERMINE THE ECOREGION OF THE SITE

The first step in ecology-based site analysis is to determine in which ecoregion the site is located. Knowing your ecoregion provides a baseline that will help you home in on the kinds of plant communities that are indigenous to the site and offers basic information about climate and soils.

This may sound daunting, but in fact finding your ecoregion involves nothing more complicated than looking at a map. If you reside in North America, you can find ecoregion maps posted online by the Western Ecology Division of the U.S. Environmental Protection Agency: epa.gov/wed/pages/ecoregions.htm. For residents of the continental United States, this site features an interactive map that allows users to click on an approximate location and open detailed regional maps with so-called Level III and Level IV ecoregion designations. The site also includes downloadable brief descriptions of Level III ecoregions, including the type of vegetation that inhabits regions at that level as well as information about climate and soils.

IDENTIFY THE HABITAT TYPE OR TYPES THAT EXIST

Careful observation of the physical environmental conditions on your site will help you determine what habitat or habitats it contains. Given that many native species grow in specific habitats, this analysis will be fundamental to most design decisions, including what and where you plant. The factors that shape the habitat on your site include sunlight, soil type, topography, hydrology, and microhabitats.

Sunlight

Light is the fuel that drives plant growth. More than any other factor, the intensity of sunlight that reaches a given area of a site determines what type of vegetation will flourish there. Gardeners have traditionally used the very crude categories of full sun, partial shade, and shade to differentiate the various degrees of light reaching their plants. Plants, however, are much more sensitive in their response to differing degrees of sunlight and the garden ecologist can benefit from following their example.

The first and most obvious sunlight pattern to note is sun versus shade, or in most cases, open meadow versus canopied woodland (although on smaller properties, the woodland may consist of just a handful of trees). Looking more deeply at the subcategories of light that occur within these two general categories will be highly beneficial.

In the open meadow, for instance, a south-facing slope will receive more intense sunlight than one that faces north. This is because the sun remains in the southern half of the heavens most of the year in the northern hemisphere. Consequently, the sun approaches a south-facing slope from a near perpendicular angle, while a north-facing slope receives sunlight from a less direct angle. Both are completely open to the sky and lacking trees or structures that cast shade, but one is likely to favor a different suite of plants than the other.

Woodland edges are another area that requires careful sunlight analysis. Sunlight is least intense in the morning and most intense at noon, when the sun is most directly overhead. The heat that it creates, however, is greatest in early afternoon. This means

Shade Intensity in Summer

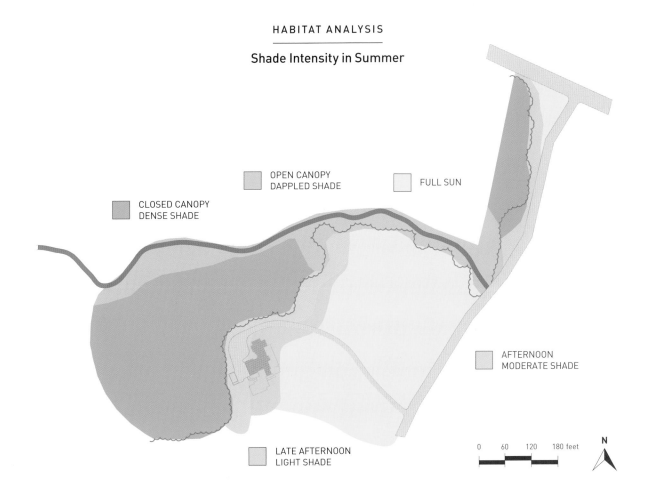

CLOSED CANOPY
DENSE SHADE

OPEN CANOPY
DAPPLED SHADE

FULL SUN

AFTERNOON
MODERATE SHADE

LATE AFTERNOON
LIGHT SHADE

0   60   120   180 feet

N

Observed sun and shade patterns can reflect habitat niches.

that an east-facing woodland edge receiving primarily morning light will support more sun-sensitive plants than a west-facing edge that receives afternoon sun. Also due to the angle of the sun, a south-facing edge may receive nearly as much light as an adjacent open field, whereas a north-facing edge may be almost as shady as the woodland interior.

Nor are all woodlands created equal, as differences in tree species, tree spacing, gaps, and overall size can create various levels of sunlight. Forests dominated by dense canopy species like maple or beech allow significantly less light to penetrate than those dominated by sparse canopy trees like oak or locust. Tree density will also affect light intensity and should be noted. Large gaps that allow in significantly more sunlight should be noted as well.

Finally, in the forest, size matters. In ecological terms, small woodland patches of less than a few acres are called fragments. From a light standpoint, these may be more similar to edge habitat as the size of the patch is too small to create a true dark interior. This effect can be particularly pronounced in long, narrow fragments, where the distance to the edge is small and light is able to penetrate throughout the fragment from the edges.

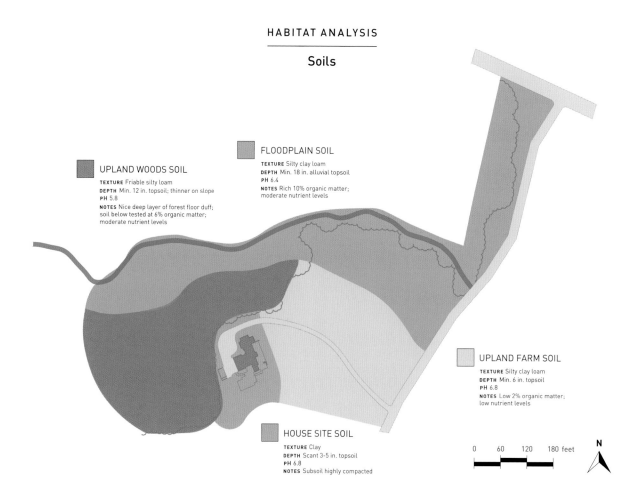

UPLAND WOODS SOIL

TEXTURE Friable silty loam
DEPTH Min. 12 in. topsoil; thinner on slope
PH 5.8
NOTES Nice deep layer of forest floor duff; soil below tested at 6% organic matter; moderate nutrient levels

FLOODPLAIN SOIL

TEXTURE Silty clay loam
DEPTH Min. 18 in. alluvial topsoil
PH 6.4
NOTES Rich 10% organic matter; moderate nutrient levels

UPLAND FARM SOIL

TEXTURE Silty clay loam
DEPTH Min. 6 in. topsoil
PH 6.8
NOTES Low 2% organic matter; low nutrient levels

HOUSE SITE SOIL

TEXTURE Clay
DEPTH Scant 3-5 in. topsoil
PH 6.8
NOTES Subsoil highly compacted

0   60   120   180 feet

N

Soil analysis of this sample site reveals rich soils in the floodplain (likely due to nutrient deposition during flooding events), lower nutrient soils in areas that had been previously farmed, compacted soils around the house, and lower pH soils in the upland woods, an area that experienced less human disturbance.

Soil type

The soil is, quite literally, the foundation of any garden, so it is essential to ascertain what you will be working with. You'll want to know the soil type (for example, clay, silt, sand, loam) and its organic content, pH, and nutrient contents. Take separate samples of the existing soil from the various sites around the property and submit them to a local soil laboratory for analysis.

The soil analysis for an ecological garden does not differ all that much from traditional horticulture. But how you use the results will. Government soil testing services were developed for the traditional brand of horticulture. With the soil test results, the laboratory will send you instructions about fertilizers and other soil amendments you should integrate to transform your soil to a good, general garden soil. This is not usually what a garden ecologist wants to do. The preferred route for our kind of gardening is to work with the existing soil. We use soil tests to tell us the existing state of our soil, rather than how to "fix" it.

There are exceptions. If a soil test reveals a serious (usually human-induced) problem with the soil, such as an abnormally high pH from past applications of lime, we will correct that. However, native species tend to be adapted to specific soil types and to compete most successfully on less nutrient-rich soils and those that haven't been amended or disturbed by extensive digging.

In addition to soil texture, pH, and percent of organic matter, soil analysis for an ecological garden should consider such factors as the degree of compaction (which may dictate the selection of compaction-tolerant plantings) and the presence near to the surface of rock ledge (which can dictate drought-tolerant plantings).

Topography

The shape of the land, its dips and rises or lack thereof, also plays a major role in defining habitat. Most importantly, it sets the drainage pattern of your landscape, determining which areas shed water deposited on them and which accumulate moisture. A depression within the landscape into which stormwater drains will be significantly moister and so suited to a different kind of vegetation as compared to the adjacent drier slopes. Even if the soil in such a depression is sandy or gravelly and fast draining, a reservoir of moisture will accumulate and remain in the subsoil, where it can be tapped by deep-rooted plants.

The transition from wet to dry can be seen by the planted species on the woodland edge. Bald cypress in the wet area at the bottom of the slope yields to upland species like northern red oak, eastern redcedar, and staghorn sumac as the elevation rises.

While moisture retention is the most obvious effect of topographical variation, there are other effects as well, including temperature. Hot air rises, and cold air sinks. Consequently low areas will not only collect moisture but also cooler air than the surrounding high ground. In addition, a south-facing slope will receive sunlight from a more direct angle than one that faces north, making it hotter and dryer. Finally, elevated areas are often exposed to more wind, which will have a significant effect on which plants will survive and thrive there.

Hydrology

Hydrology is very important, as it indicates how much moisture is available to plants. Many of the characteristics described previously can help to determine the hydrologic characteristics of various areas within the site, particularly topographical position and soil type. A vegetative analysis will also be quite helpful, as the presence of wetland plants may be the best indication that moisture levels are high in that area.

top This dry slope supports splitbeard bluestem (*Andropogon ternarius*), a beautiful but little-used native grass.

bottom Goldenrod and mountainmint are well suited to moderately dry land.

Water table depth is another characteristic that is important to note. This can generally be determined by taking a deep core soil sample and examining whether mucky soil exists, or whether the hole that remains after removing the soil fills with water. If either mucky soil or water is present, the depth where that begins should be noted. A soil core sampler, a simple hand tool that acts like a corkscrew to remove a deep cylindrical core of soil, can be used for this. In addition, high water tables are often a regional condition and can be determined through local ecological literature or discussion with a local ecologist.

The presence of springs, which generally affect limited areas on the site, are also important. To determine if these exist and where they may be located, look for isolated pockets of wetland vegetation or areas where the existing vegetation is larger and greener. The latter can even be observed in turf areas where native vegetation is not generally present as a result of mowing. Springs generally affect not only the area where the water nears the surface, but also the area into which it drains. Consequently, follow that route and notate where it ultimately collects.

Areas where significantly more moisture exists than the surrounding areas, on various scales, are referred to as wetlands. Determining where these exist is obviously crucial to the selection and placement of plants, but it is equally important to distinguish between

Topography and Hydrology

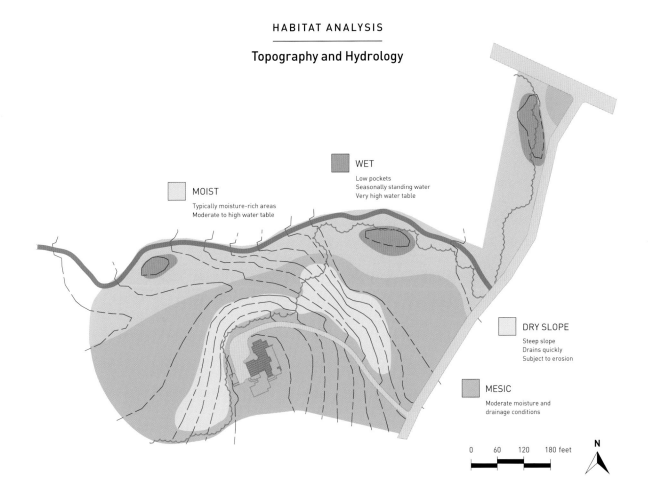

WET

Low pockets
Seasonally standing water
Very high water table

MOIST

Typically moisture-rich areas
Moderate to high water table

DRY SLOPE

Steep slope
Drains quickly
Subject to erosion

MESIC

Moderate moisture and
drainage conditions

0   60   120   180 feet

N

Careful observation of a site's hydrology and the relationship to topography can reveal a variety of microhabitats, from dry steep slopes that drain quickly to wet low pockets with seasonally standing water.

two types of wetlands: permanent wetlands, which always remain wet, and ephemeral wetlands, which are inundated with moisture during rainy periods but dry out at other times. Different plant species tend to occupy areas where each of these conditions exist. Those that grow in permanent wetlands are referred to as obligate wetland species, whereas those that occupy ephemeral wetlands are referred to as facultative wetland species. Familiarity with these terms is important, because many native plant nurseries use these terms to describe the plants that they offer.

In your planting program, it is easy to ignore the fine points of orientation, exposure, and topography that define existing habitats, but you will find that the plants do not. Consequently, it is far more efficient and economical to spend some time reading the site before planting and then placing the different species where they want to be.

Microhabitats

There are invariably spots within larger habitats where some particular feature creates a small variation in the ecological condition as it relates to plant growth. Examples include the shade cast by a lone tree in a field that is largely sunny, or a small low pocket in the landscape where more moisture collects than the surrounding area.

Keep an eye out for such anomalies as you acquaint yourself with the site, as microhabitats can provide opportunities if skillfully exploited. A stonewall located on a slope will create a cold pocket at its north face, as it traps cold air that tends to sink and travel downhill. Consequently, a gardener in Pennsylvania may be able to grow plants that are native farther north, such as North America's only native yew, Canada yew (*Taxus canadensis*). If this cool microhabitat also happens to be located in a wet woodland area, the same Pennsylvanian may be able to plant the beautiful pink-flowered rhodora (*Rhododendron canadense*), a rare and delicate rhododendron found in northern New England and eastern Canada.

If ignored and treated the same as the surrounding area, microhabitats can be troublesome: the planting that thrives on the better-drained surrounding area is likely to die if dropped into a low-lying, wet spot. That's why, for a garden ecologist, working with existing conditions, including their minute variations, indicates what species and type of community a site will naturally favor, rather than creating a recipe for improvement.

INVENTORY EXISTING VEGETATION

If you could perform only one site analysis task, taking an inventory of the unplanted vegetation on your property would probably be the best choice. Regardless how carefully you note the habitat characteristics described above and subsequently try to match the right plant with the right place, the plants themselves can do it better. If a native plant established and grew in a wild spot without any human assistance, that plant will likely thrive with minimal human assistance in a nearby garden containing similar conditions. The effort involved in completing such an inventory can even be valuable on a property where there are only a few native species to be observed. A single, straggly white turtlehead (*Chelone glabra*) in a forgotten unmowed corner of the property can tell you that this species is likely to grow well here or in similar areas of your property, and you should consider including it in your planting plan. Likewise, the entire wet meadow plant community that grows with white turtlehead, including monkeyflower (*Mimulus ringens*) and Virginia mountainmint (*Pycnanthemum virginianum*), is likely to grow well here or in similar areas of your property, and you should also consider including them in your planting plan. Thus, observing one plant in a tiny patch can provide you with an entire plant community on which to model your plant palette for particular areas.

In additional to identifying native and other desirable species, it is essential to identify any invasive plants on the site. Otherwise, they may overrun any openings you create in your design and overwhelm any new plantings. Identifying undesirable species will allow you to formulate a long-term plan for their control before you begin any new development of the site and avoid a series of long-term and possibly futile battles.

It's important to recognize, however, that not all plants of foreign origin are problematic. Chicory (*Cichorium intybus*), for example, is a European perennial that has integrated itself into the flora throughout North America south of the Arctic without becoming invasive and displacing native competitors. On the other hand, certain very aggressive natives also bear watching. Canada goldenrod (*Solidago canadensis*), for instance, earns its credentials as a native by flourishing throughout most of Canada and the United States, except for the Gulf South and the extreme southeast, yet the aggressiveness with which this rhizomatous perennial spreads has made it a liability in domesticated landscapes. Fortunately, there are many other species of goldenrods with similar yellow flower sprays that don't crowd out their neighbors.

A thorough inventory should be performed more than once during the growing season, as plants that are not actively growing at the time of the inventory may be missed. A spring study may not reveal the presence of warm-season grasses like little bluestem, as they do not actively grow until the onset of warm weather. Similarly, a summer inventory may miss the presence of spring ephemeral woodland wildflowers, as they lose their leaves after flowering in the spring.

If you are lucky and the human disturbance to your site has been minimal or even moderate, there are likely to be remnants or at least hints of a natural plant community persisting there to observe. Even if humans have had heavy impacts on the site, clues to the earlier vegetation may survive. Look to unmowed corners or the out-of-the-way edges of the property, where cultivation has been less thorough, and you may find remnant native plants that suggest what plant community flourished there before the site was disturbed.

Of course, much or all of the surface area on many suburban and urban properties is covered by turf, weeded gardens, asphalt, and a house. If you live in such an area and your site has been too thoroughly disturbed to host any native remnants, then the best means of obtaining information about what was likely growing there previously is to observe the vegetation on nearby properties where unmanaged areas do exist. These can include local nature centers, arboreta, or private properties. Local nature centers and arboreta can also provide introductions to experienced members who can share information about what they have seen and grown in your area. In addition, such organizations commonly sponsor field trips, which offer a novice the opportunity to discover local habitats and plant communities in the company of more educated members.

After finding useful areas to inventory, plant identification is your next job. A good old field guide is still a useful way to accomplish this, but numerous websites and apps for mobile devices are now becoming available to make this task even easier.

Whatever you find growing on your site, chances are good that it hosts far more potential plants than the visual evidence suggests, in the form of dormant seeds in the topsoil wherever it has not been dug and disturbed. Such sleepers comprise the seed bank of the site, and seeds can retain their viability for years or even decades, waiting to sprout when conditions become favorable. The seed bank has a great influence on which unplanted species the gardener will eventually deal with, good and bad, yet its presence is infrequently considered in traditional horticulture.

When you turn over the soil for a new garden, remove the turf, or rip out the vines in the woods, conditions become favorable for the seed bank. Resulting soil disturbance or even just the removal of competition will initiate this process. This can be intentionally performed in small test plots to determine what the seed bank's response will be to the actions you'll take in preparation for planting. In some cases, the resulting vegetation may

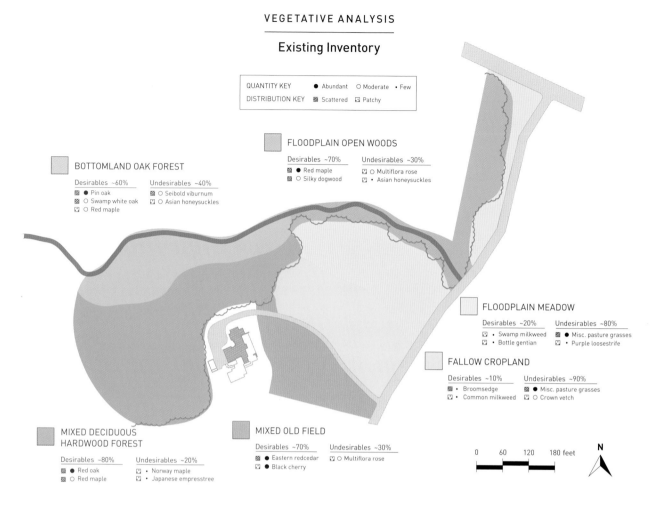

Existing Inventory

QUANTITY KEY	● Abundant	○ Moderate	• Few
DISTRIBUTION KEY	▨ Scattered	▱ Patchy	

FLOODPLAIN OPEN WOODS

Desirables ~70%	Undesirables ~30%
▨ ● Red maple	▱ ○ Multiflora rose
▨ ○ Silky dogwood	▱ • Asian honeysuckles

BOTTOMLAND OAK FOREST

Desirables ~60%	Undesirables ~40%
▨ ● Pin oak	▱ ○ Seibold viburnum
▨ ○ Swamp white oak	▱ ○ Asian honeysuckles
▱ ○ Red maple	

FLOODPLAIN MEADOW

Desirables ~20%	Undesirables ~80%
▱ • Swamp milkweed	▨ ● Misc. pasture grasses
▱ • Bottle gentian	▱ • Purple loosestrife

FALLOW CROPLAND

Desirables ~10%	Undesirables ~90%
▨ • Broomsedge	▨ ● Misc. pasture grasses
▱ • Common milkweed	▱ ○ Crown vetch

MIXED DECIDUOUS HARDWOOD FOREST

Desirables ~80%	Undesirables ~20%
▨ ● Red oak	▱ • Norway maple
▨ ○ Red maple	▱ • Japanese empresstree

MIXED OLD FIELD

Desirables ~70%	Undesirables ~30%
▨ ● Eastern redcedar	▱ ○ Multiflora rose
▱ ● Black cherry	

0   60   120   180 feet

N

Methodical inventory of existing vegetation, both desirable and undesirable, and the quantity and distribution of existing growth can reveal a lot about a site, its history, and its potential futures. This example vegetation inventory defines major vegetation zones with a few species listed for each; your list will likely be much more comprehensive.

be more of what you already observed. Sometimes, however, new species will show up, giving you further guidance on what native plants are adapted to your site, as well as which plants are likely to show up on their own going forward. Testing the seed bank composition can also be performed indoors, under controlled conditions.

DETERMINE THE VEGETATIVE TRAJECTORY OF THE SITE AND WHERE YOU ARE ON THAT TRAJECTORY

Be aware as you conduct your site analysis that because the vegetation continually evolves, the wild plant compositions you observe today were probably quite different in the past. An upland grassland, for example, if left undisturbed, is likely to be replaced by the shrubs and trees of an oldfield plant community, which, in the eastern half of the United States, the upper Midwest, and the Pacific Northwest typically gives way to pioneer woodland and then a mature woodland. In coastal southern California, grasses and annuals may transition to a relatively stable shrub composition known as chaparral, while in some arid climates of the Southwest the evolution may tend toward a shortgrass prairie and never succeed to shrubs or trees. In fact, your task is not just to identify the plants that currently occupy your site, but to determine what stage of natural succession the site is currently in and what stages likely preceded it. If you understand the natural progression of vegetation in your region, you will be able to determine where in this timeline the various areas of your property currently fall, which will allow you, in effect, to predict the future.

These processes of vegetative change would likely occur without human management or interference, but virtually everywhere we garden has been subjected to human alteration, usually in spades. Reconstructing the history of logging and agriculture on your site, as well as more recent construction activities like clearing or grading, will be more than an interesting research endeavor. Which of these disturbances have or have not occurred will also have a huge impact on how vegetation will play out as you implement your plans. Possibly the most important aspect to determine is not *whether* human disturbance occurred—it has occurred virtually everywhere—but what *type* of disturbance occurred.

Reading this history is a matter of learning to recognize the clues. A line of larger trees in the woods may be the remnant of a hedgerow that once marked the edge of a field. A stone wall also testifies to an agricultural past and suggests that the field had been cultivated, as it is only necessary to remove stones if tilling is to be performed. A remnant hedgerow without an associated stonewall may indicate that the field was only grazed and never cultivated, particularly if significant stones or boulders remain in the current or abandoned field.

Particularly significant is a single tall and spreading tree in the middle of an open field or among young spindly pioneer trees in the woods; this is undoubtedly a surviving wolf tree, a loner that a farmer allowed to remain in the pasture to provide midday shade to cattle. The presence of such a tree strongly suggests that the area was used for grazing. Determining whether a field historically was cultivated or grazed is actually quite relevant to your efforts. Grazed pastures, particularly those that were too wet, too ledgy, or too steep to plow, may have never been subjected to soil disturbance. This means that the soil profile and seed bank composition is largely intact. These areas tend to contain a higher proportion of native species than areas where soils have been disturbed, and they can be surprisingly resistant to invasive species encroachment. Consequently your garden or restoration efforts in this type of area will likely be easier and result in greater plant diversity and ecological complexity. Or, translated into normal language: it will look better.

There are other clues besides remnant hedgerows and stone walls that can indicate the disturbance history of a property. First, if an area has physical impediments to plowing, such as steep slopes, surface ledge, or excessive moisture, it probably wasn't. In areas where farming is still practiced, there are often neighbors who can recall the relatively recent land-use history. Sometimes, however, the plants can reveal the story. Land on which naturally recruited native plants dominate and have resisted invasive species incursion often has less soil disturbance in its history or possibly none. If a field or sunny opening has a dominant cover of little bluestem, asters (*Symphyotrichum* and *Eurybia* species), staghorn sumac (*Rhus typhina*), and a few stray butterflyweeds, it is likely that this is a natural occurrence due to lack of disturbance. A similar conclusion can be made about a woodland ground layer where mature woodland herbs like bloodroot, largeflower bellwort (*Uvularia grandiflora*), or trilliums are found. The presence of these plants tells you that these late-stage and more difficult to grow species can be successfully introduced. In areas where these mature forest herbs are not present, it will be better to plant generalist species like white wood aster or wreath goldenrod (*Solidago caesia*).

Where possible, it is also important to map where more recent disturbance occurred on your property. Most commonly this would have occurred during construction activities.

Analyzing where in the process of succession the existing naturally occurring plant compositions fall and how human disturbance has affected those compositions will be crucial in determining whether ecological conditions are present to support late-stage specialist species and, if so, where. That same disturbance history can also tell you how extensive invasive plant pressure is likely to be. And if you have determined that all or part of your project area has never been cultivated, plowed, or bulldozed, by all means, do *not* disturb!

ANALYZE THE PRESENCE OF WILDLIFE

A key goal of ecological gardens is to create an attractive environment for wildlife. While native plantings can contribute significantly in this regard, you also want to make sure that none of your activities dissuade the species that are already there. The wildlife species are on your property because the resources they need to survive are on your property. Consequently, you want to avoid removing existing native species that wildlife may be utilizing.

In addition to preserving individual native plants, the presence of a diversity of plant community types, including open field, shrub thicket, and woodland, may be supporting existing wildlife, and preservation should be considered where possible. Even invasive species, such as multiflora rose (*Rosa multiflora*) or common buckthorn (*Rhamnus cathartica*), may be providing protective thickets. If so, they should be transitioned over time into native shrub thicket that will provide as good or better cover, without leaving the area denuded for an extended period of time.

Because we live in a world where balanced ecosystems of interdependent plant and animal species have been highly disrupted, some wildlife—both native and imported—can cause significant problems in naturally occurring and planted landscapes. The rapacious appetite of white-tailed deer is well known, and it is of course crucial to analyze their population levels when designing a landscape. If you have lived on your property for a while, you are well aware whether deer are present in high numbers. If you have recently acquired your property, a quick conversation with one of your new neighbors will likely clarify the situation.

If the neighbors' litany of deer-related complaints, or lack thereof, isn't sufficient, there are other ways to determine the presence of deer. Scout for droppings or evidence of plant predation in existing gardens to confirm their presence. If highly favored plants like yew, hosta, or rhododendron are thriving in the landscape, lush and fully foliated, deer visitation is probably low. Observing ground-layer vegetation in the woods or its lack thereof can be particularly helpful. A healthy woodland will always have a crop of tree seedlings and saplings present to replace the aging trees in the canopy. The absence of these young trees generally means heavy deer predation. The shrub composition can also be a good indicator. If only unpalatable shrubs or understory trees like northern spicebush (*Lindera benzoin*) or pawpaw (*Asimina triloba*) are present, deer are likely overpopulated in your area. Deer browse can also be seasonal, with heavier pressure at certain times of year when food is less available in nearby natural areas.

Canada geese are another wildlife species that can be a problem when large numbers inhabit a landscape. These birds may pull newly planted perennial plugs out of the ground, seemingly out of spite, as they often don't eat these victims but leave them lying. Geese droppings also add high levels of nitrogen to the soil and contribute to eutrophication of waterways. As they wander lawn areas in large flocks, the blanket of droppings they leave and their generally cantankerous presence can make things quite unpleasant for people.

In grasslands and pastures, be alert for the presence of grassland nesting birds. Consider them when deciding when to mow, and inspect the meadow for nests before doing so.

Planting vegetation that is above their eyesight along the perimeters of the turf areas, particularly at the water's edge, will discourage the presence of geese as they do not like to pass through vegetation that they can't see predators over. Noting the presence of geese and how they are travelling on the property can allow you to place vegetation in a manner that discourages them.

As if geese and deer are not enough, the colonization of invasive earthworms can also have an adverse effect on vegetation, particularly in woodlands. These exotic earthworms, some accidentally introduced as worms sold for fishing bait, have escaped into the wild, where they voraciously consume almost all of the organic matter in the soil, making it very difficult to grow plants. Their presence can be detected by examining the soil, which will be extremely light in weight and seem like it is composed of hollow casings. Identifying this

The step placement and surrounding plantings were designed to highlight a view of an attractive tree trunk arrangement.

problem, if it exists, will allow you to select plants that are more likely to tolerate these difficult conditions, such as the native woodland stonecrop (*Sedum ternatum*) or wild ginger (*Asarum canadense*). It can also indicate the need for more frequent additions of organic material than is normally needed in unaffected woodlands.

The analysis described above looks at the general environmental conditions that could be providing good habitat for wildlife on your property. You may, however, prefer to take the wildlife aspect of your property to an even deeper level. Often the desire to attract a wide diversity of bird species, including some less common ones, drives this interest. If this is the case, it is advisable to inventory the particular bird species present. This can be a difficult endeavor for a novice as some birds may be hard to identify or even locate. Extensive work with a field guide may be necessary, or it may be helpful to bring in a wildlife ecologist to provide an inventory. Species have specific requirements. You can only make sure that they remain if you understand what aspects of your landscape are keeping them there and needs to be preserved. In addition, an ecologist may be able to advise you as to what species have the potential to be there, if certain elements were added.

Views

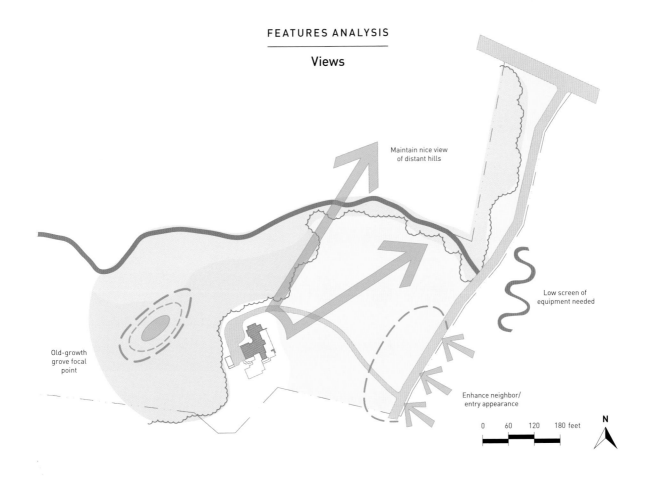

Maintain nice view
of distant hills

Low screen of
equipment needed

Old-growth
grove focal
point

Enhance neighbor/
entry appearance

0   60   120   180 feet

N

When assessing a site's existing aesthetic features, take into account special vegetation, on- and off-property views, and areas to enhance or screen.

IDENTIFY THE SITE'S EXISTING AESTHETIC ATTRIBUTES, LIABILITIES, AND LIMITATIONS

Sometimes the best aesthetic features of a completed garden are not those we created but those that we find already in place. Indeed, a fundamental role of the designer is to find and accent such exceptional existing features. If the site contains a beautiful grove of trees or even just one exceptional tree, note that so that you can expose it and make it a focal point of your garden design. Trees that have rejuvenated from previous storm breaks often contain architectural twists and turns, which can be revealed by removing surrounding vegetation or enhanced by artistic pruning. Natural stone outcrops, topographical varia-tions, attractive vistas, striking plant compositions, and a host of other elements should be carefully noted and located. An unconventional but potentially very attractive feature is a snag, a standing dead tree. If this corpse is large and sculptural, it can lend atmosphere to the garden as well as providing a haven for woodpeckers and other wildlife.

Off-property views, both good and bad, should be identified so that they can be used or screened as necessary. The same view can be both good and bad; for example, you may wish to use a low screen to block the view of the road, while leaving the view of the

This planting was designed for aesthetics, wildlife habitat, and to screen views of the mowed lawn next door.

woodland beyond it unimpeded. Consequently, it is important to note not only where screening is needed but how high it should be to screen only the undesirable element(s). This can be done by having one person stand at the source of the view, while another stands at the point that a screen planting is needed. By raising a pole to the point where the element to be screened first disappears from the viewer, you can determine the optimal height for your selective screen.

The aesthetic character of your designed landscape, particularly the level of wildness that you intend to achieve, will inevitably be affected by surroundings. A relatively ungroomed wild woodland or meadow garden will look at home in an uncultivated setting, or one in which neighbors have accepted the natural character of the landscape. In a neighborhood of neat lawns and flower beds, the same garden will make a very strong statement, so you may wish to plan a somewhat more groomed design.

In almost every landscape setting, there is something existing that is worth revealing or highlighting. Even a newly created suburban lot, where the bulldozers have left nothing but a flat landscape with lawn and a minimal foundation planting, will often have views in the distance of woodlands or tree groves that can be selectively featured by creating windows in the plantings that screen you from the rest of the neighborhood.

CREATE A BASE MAP SUMMARIZING ALL THE ATTRIBUTES

It is important to note not only what ecological conditions occur on your site, but where they occur. Mapping these patterns on paper, or at least noting where all of the differing conditions occur, will create a habitat and natural features base map. The subsequent planting or garden plan can essentially be an overlay of this map, allowing you to dovetail as much as possible the ecological and other criteria on which you are basing your design.

OTHER RESOURCES

Collecting and interpreting information on the ecological conditions of your site in the detail that I have described may seem daunting. But if you are a homeowner, you need only learn about one place: yours. Many resources are available, including books, online sources, and ecologists and botanists at local nature centers and arboreta. More than likely they will be eager to help, as public education, particularly on ecological matters, is often part of their mission. If you are a professional and need to work over a wider geographical range, a more serious study will be well worth your while, starting with membership and participation with groups involved in ecological restoration, such as the Natural Areas Association, The Society for Ecological Restoration, and the Lady Bird Johnson Wildflower Research Center. Either way, the more you understand where you are, ecologically speaking, the more you will be able to influence the path your garden will take.

During site analysis, watch for site features to highlight, such as this prominent ledge and existing vegetation. Here, soil manipulations transformed small patches of existing moss to an all-encompassing carpet.

REVERSING SUCCESSION

Sometimes, when you are very lucky, the best results require relatively little effort. That was the case in a meadow I created in Dutchess County, New York. In fact, my greatest contribution to the success of this project lay in what I didn't do.

The site was on an extensive property. The house was a historic recreation, and the garden immediately surrounding this was a carefully orchestrated display of beds and borders. I was hired to address a large woodland opening a short walk from the house, a full 14 acres in size. It was an example of advanced oldfield succession, with relatively few herbaceous plants and perhaps 95 percent of the area covered with gray dogwood (*Cornus racemosa*). This is an attractive, if large, shrub that reaches a height and spread of 15 feet and bears bunches of white flowers at the twig tips in late spring followed by small white berries. The leaves turn purplish red in autumn, dropping away to reveal colorful red stems. But the owners of the property wanted an open place to stroll and enjoy vistas, not a dense brush patch or more woods, which was what, if left alone, this opening would turn into in a few years.

The obvious response was to kill and remove the shrubs and plant a meadow. Fortunately, I had closely examined the site and found in the small surviving areas of herbaceous plants some remnants of little bluestem, one of North America's most attractive native meadow grasses. Prized for its blue summer foliage, fluffy silver flowers, and orange autumn and winter color, its presence suggested that prior to the proliferation of the gray dogwood this area had been a little bluestem meadow. The presence of this native grass also

suggested that this area, back when it had been some farmer's field, had never been plowed but was more likely used for grazing. Plowing would have eliminated any native remnants, including the little bluestem. When I checked the soil, I found that it was shallow, with rocky ledge close to the surface in many spots, which would have made this area unsuitable for plowing and cultivation. This reinforced my belief that the only human disturbance to the meadow vegetation on this site had been the pasturing of grazing animals.

I didn't want to remove all the gray dogwood, which in addition to its aesthetic appeal provides good cover for birds, as well as material for framing views and creating focal points. The goal became rolling back the successional clock to a very specific point in the field's past when the cover had been less dogwood—perhaps 15 percent—and more little bluestem and associated wildflowers. I chose carefully scheduled mowing to accomplish this.

I marked the areas where I wanted to preserve the dogwood, and then had the rest of the area brush-hogged, a low-impact process. One elderly man accomplished the brush-hogging with an aged Gravely walk-behind tractor. I could follow his progress through the brush by the sound of the mower interspersed with clunks, followed by loud profanity as the tractor found another of the rocks that had saved this field from plowing. I could feel his pain, as I'd owned one of these tractors myself many years back. While these rugged machines could probably cut down a telephone pole, they were also known as "man killers," and my bruised hands and sore muscles had prevented me from ever cozying up to them.

Observing existing vegetation and using simple management procedures enabled the conversion of an impenetrable shrub thicket into a sweeping expanse of little bluestem, one of North America's most beautiful native grasses.

The timing was critical. The site was mowed repeatedly in the spring, when shrubs such as gray dogwood normally achieve most of their annual growth but when the little bluestem, a warm-season grass, is still dormant. My team stopped mowing with the onset of summer, when the shrubs stopped growing and the little bluestem took off. By repeatedly removing the new growth from the dogwoods in this fashion, we starved them over a period of a few years. Meanwhile, the little bluestem, which was never being cut during its growth period, was free to photosynthesize and produce food to its maximum potential. Late-emerging wildflowers like asters and goldenrods were also favored by this process, and after four years we had a dense little bluestem meadow. The mow regime was then reduced to once yearly in late winter.

The only variation in this expanse of grassland was periodic gray dogwood drifts that had been excluded from the mow regime. These preserved drifts were carefully located in the meadow interior to gracefully guide visitors through the field, and on the surrounding woodland edges to frame particularly good views into the forest.

But given that these drifts were not being cut or deprived of their ability to produce food, how would we keep these aggressively spreading shrubs from expanding and recolonizing the meadow? A series of walking paths that meander through the meadow were also designed to encircle each dogwood patch. As these paths were mowed, the new shoots were continually cut off as they emerged from the shrubs' expanding root system, holding the patch in place. From an ecological succession standpoint, our initial spring mow program had turned back the hands of time, and our dual-purpose path system was allowing time to stand still.

All this was accomplished without any tilling, spraying, or planting. From a physical standpoint, our elderly friend and his Gravely mower had single-handedly transformed an impenetrable thicket into an intricate

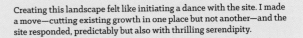

Creating this landscape felt like initiating a dance with the site. I made a move—cutting existing growth in one place but not another—and the site responded, predictably but also with thrilling serendipity.

native meadow. From a mental standpoint, this was made possible through a plan of action that combined the thoughtful manipulation of ecological succession with garden aesthetics. And the aesthetic attributes of the composition were not only derived from the beauty of the individual plants but also from the graceful interplay of our natural landscape's basic building blocks: meadow, shrubland, and forest.

Had this field been cultivated and plowed in the past, it would have lacked the native grass remnants and I couldn't have returned it to meadow so easily. Had I followed standard horticultural procedure and pulled or dug out the shrubs and then tilled the soil for planting, I would have also destroyed this field's inherent potential. Instead of promoting the re-emergence of the meadow, I would have had to start a meadow from scratch. (Ironically, I might well have planted little bluestem, because it is such a superior meadow grass for the Northeast.) In addition, the resulting soil disturbance from tilling would have opened the site to opportunistic weeds, and made the meadow much more difficult to manage going forward.

My choice of the easier path in this case was not fortuitous. It was the result of a careful reading of the site and taking advantage of what the site taught me about its own ecology. This technique would not have worked had the farmer's field been cultivated for raising crops. It is as important to know when you cannot use a particular technique as when you can, and it is the site that tells you this either way. Had I not paid attention, I might have committed a truly oafish act. The loggers who had cleared the land hadn't disturbed the soil, nor had the farmer who worked it later. But I, in a misguided desire to return the field to "natural" vegetation, might have been the first person in the history of human interactions with the site to till the soil and change it irreversibly. The time I invested in inspecting and analyzing were truly hours well spent.

CREATING AN ECOLOGICALLY CONNECTED MASTER PLAN

You've traced the history of your property and analyzed its ecological state. Now you can begin to envision the types of ecosystems and landscape compositions it can foster going forward.

First, think big picture. An architect selects an overarching architectural style long before considering the stencil patterns for the kitchen walls. In a landscape, it is also best to formulate an overall vision before choosing the individual plants. This vision must be based on the inherent ecological patterns and processes of your property, the factors that you worked out on the habitat and natural features base map that was the culmination of the site analysis process described in the last chapter.

Paths can enable a journey through openness and enclosure. Here, a stepping-stone path leads through dense shrubs to a terrace located in a woodland opening.

PRESERVING YOUR SITE'S ECOLOGICAL ASSETS

Like a physician, the garden ecologist should take "first, do no harm" as a guiding principle. Protecting existing native growth, particularly woodlands, is easier and less expensive than trying to restore it after it has been destroyed. Even our best restoration efforts may never achieve the beauty and mystery of undisturbed forest, where ferns cascade over rocks and wildflowers spring up under a canopy of interlaced branches.

White wood aster is a generalist species found in most woodlands in the eastern United States. Although this species should be preserved where possible, protecting it represents a lesser priority than protecting those species that are more specialized.

Early decisions relating to the siting of buildings, topographic changes, and excavation disturbance can help minimize destruction of natural growth during construction. Unfortunately, landscape design considerations are often delayed until after construction of the home or other structures is complete, and by that point, the damage is done in most cases. The earlier you can begin envisioning the directions you'd like the landscape to take and begin protecting its existing elements, the better.

Consideration of ecological impacts should play a role not only in the placement of major constructed elements, such as the house or driveway, but also in garden elements as well. When siting a pergola, for example, traditional gardeners may content themselves with considering where the structure will be most accessible and where it will sit most gracefully in the landscape. A garden ecologist, however, will also seek to place such a feature where it will not disrupt an existing ecological asset or preclude a potential one. From an ecological perspective, you should ideally place the pergola where weedy growth dominates and little of environmental value will be disturbed by the construction; in fact, such weedy sites are usually the result of some previous disturbance. If an appropriate weedy area is not available, you could locate the structure in an area dominated by common generalist native species, particularly if those species enjoy substantial populations elsewhere on the property. But where highly specialized native plants are present, particularly if they so far have resisted invasion: do *not* disturb! Not only will construction of the pergola displace these desirable plants, this work will tear openings in the vegetative fabric that will enable invasive species to colonize what was previously a stable, native plant population.

Two examples of highly evolved plant associations in the United States are the northern hardwood-conifer forest understory specialist association, which includes blue cohosh (*Caulophyllum thalictroides*), Canada mayflower (*Maianthemum canadense*), Dutchman's breeches (*Dicentra cucullaria*), ferns, partridgeberry (*Mitchella repens*), clubmoss (*Lycopodium* species), trilliums, and spring ephemerals, and the tallgrass prairie specialist association, including big bluestem (*Andropogon gerardii*), compassplant (*Silphium laciniatum*), blazing star (*Liatris* species), wild indigo (*Baptisia* species), phlox (*Phlox* species), prairie clovers (*Dalea* species), and wood lily (*Lilium philadelphicum*). Additional plant associations can be found by consulting ecoregion and plant community references for your area.

In addition to noting where specialist plant communities exist on the site, your base map will show areas where there is the potential to establish uncommon, specialist plants. For example, a low area in the lawn that is inundated during rainy periods but dries out when there is little rain may support an ephemeral wetland plant community, which thrives in such conditions. If dry upland conditions occupy most of the property, a spot of this kind may offer an opportunity to diversify the plant palette for wildlife and aesthetic appeal.

Avoid disturbing woodland areas where highly specialized species, such as Dutchman's breeches, are present.

A traditional gardener might say, "Fill in that useless soggy area and put the pergola there." A garden ecologist is more likely to say, "Don't put it there. That's the only place I can grow pink milkweed and cardinalflower." The wetland garden can be designed with an adjacent sitting area, so that you can enjoy the butterflies and hummingbirds that the wetland vegetation will attract.

ARRANGING SUCCESSIONAL ZONES

In creating the natural features base map, you have located the various physical habitats on the property. Later in the design process, you'll identify the best adapted native plant community for each of those habitats, using this as a guide for plant selection. In addition, however, you must also choose at which stage of the successional process you want to fix each habitat. In much of North America, such stages consist of herbaceous meadow, shrubland, and forest.

As part of your analysis of the landscape, you noted the distribution of existing successional zones. Now, in the design process, you are likely to be adding to, extending, or even subtracting from them as you work to make the landscape not only ecologically healthy and productive but also pleasing and practical. What types of vegetation do you intend to enhance or create, and where will you locate the boundaries of each? Will there be a woodland? A shrubland, hedgerow, or meadow? How will you arrange and configure these?

When working on this, I find it useful to think in terms of what noted landscape designer Darrel Morrison describes as a "mass/void plan." That is, how will you balance the imposing mass of woodland or shrubland against the open space of meadow? As you work this out, plan, too, how each area will transition into its neighbors.

Take the time to work this out on paper, perhaps on tracing paper in an overlay for your habitat and natural features base map. The way in which you arrange the successional zones will have an important effect on how the different habitats interact and function. Factors that should affect the arrangement of successional zones and the resulting functioning of habitats include area size and shape, connectivity between habitat patches, transition zones, and the aesthetics of the landscape pattern.

AREA SIZE

From an ecological perspective, a habitat patch of 1 acre, 5 acres, or 50 acres all act quite differently, even if they contain the same exact plant species. For example, many woodland birds, particularly some warblers and other songbirds, will not nest in a forest patch that is smaller than 50 acres. Many ground-nesting meadow species, such as bobolinks and meadowlarks, will similarly only nest in large tracts of their preferred open meadow habitat.

Plants also respond to area size. This is particularly true in woodlands, where light increases as you approach the edge, with the result that sun-loving vines and other invasive species often dominate there. Conversely, darker woodland interiors favor the native shade-loving species that are better adapted to lower light levels. Larger woodlands have a higher ratio of interior to edge and thus more low-light habitat, which favors native ground-layer species.

Given that native wildlife and plants will generally be more successful in larger patches of their preferred habitat, it is important to consider the size of each in your plan. Planting an expansion of existing woodland will generally have more value than a new planting of the same size that is freestanding. Also, planting three 1-acre patches of woodland will have much less wildlife value than creating one 3-acre patch.

In another instance, you may want to remove some woodland to accommodate a native meadow. Although cutting down trees sounds like a very un-ecological thing to do, in many areas the bird species that are declining most rapidly are those that use meadow and grassland habitats, so this change may actually be beneficial, assuming the created meadow is large enough to support grassland bird species of conservation concern. If you do decide to remove some trees, keep in mind that the ecological value of forest is not only determined by how much forest you leave but also where you leave it. Splitting a 3-acre woodland in half by inserting a 1-acre meadow completely across the middle leaves two

1-acre patches of woodland. Putting the meadow on either end or carving out a U-shaped section that leaves a portion of the woods connected leaves one 2-acre woodland patch. In both cases the woodland area totals 2 acres, but the larger configuration provides much better habitat for the plants and wildlife you want to encourage.

AREA SHAPE

In addition to an area's size, its shape also influences habitat value and plant colonization. Many wildlife species, particularly birds, require minimum areas of their preferred habitat. As many predators live or hunt at or near the edges of habitat patches, be it woods or open field, the farther the potential meal is from the edge, the safer that target will be. Similarly with plants, a larger habitat size and one with an unattenuated shape allows them to better proliferate, as they can be positioned farther from the edges where increased competition exists from aggressive invasive species. Size of habitat matters, but so does shape.

To understand the importance of shape, compare two 4-acre habitat patches that are configured differently, one being a square and the other a long, narrow rectangle. A plant or animal in the center of the square is much farther from an edge than one occupying the narrow rectangle.

Of course, you needn't limit yourself to squares and rectangles. The same ecological effect can be obtained through wavy edges and asymmetrical shapes. Irregular edges, where an interlocking interplay between woods, shrubland, and meadow can work together, create a dynamic experience as one traverses the landscape and views it from different perspectives. The ecological benefits of such interactions are likely to be considerable, as there are distinct classes of plants and wildlife that prefer to inhabit such ecotones.

CONNECTIVITY BETWEEN HABITAT PATCHES

Disconnected patches of woodland, shrubland, or open field may already exist on your property. Alternatively, practical or aesthetic aspects of the plan may prevent you from maximizing the size of the individual areas of habitat types you are planting. Even species that can use the resulting small spaces may not want to travel through the dissimilar habitat that separates the fragments, especially if they are separated by lawn that offers no protective cover. Planting narrow connecting corridors with similar vegetation can allow the fauna to access both areas, while not using up space that may be needed for other elements of the plan. Even though the corridor may not be wide enough to provide the animals in question with much habitat value in and of itself, it enables them to access both fragments, effectively increasing the area where they can seek food and cover from predators.

What's more, when animals move through such corridors, so do plants. Species whose seeds are disbursed by sticking to or being eaten and defecated by wildlife can easily hitch a ride from one patch of habitat to the next. Even plants that are dispersed by gravity, expanding their colonies slowly as seeds fall and germinate near the mother plant, will benefit, albeit slowly, as their expansion will not be blocked by area in which they cannot grow.

In addition to ecological enhancement, habitat corridors can offer aesthetic and practical benefits. A drift of meadow winding across a large area of turf connects adjacent meadow fragments while enhancing the visual dynamic of the previously unbroken lawn. A mass of shrubs that connects existing but isolated shrub fragments can be designed with an opening in the middle to create an intimate and private getaway.

Screening is a common practical need in landscape design, and this can lend itself to the creation of habitat corridors. In traditional design, a screen typically consists of a single-species row of evergreens such as arborvitae. A garden ecologist is more likely to replace that file of arborvitae with a mixed border of evergreen shrubs and trees that mimics a native hedgerow. In addition to blocking unwanted views, this latter type of planting can serve to connect isolated woodland patches on the property. In fact, if similar native hedgerow compositions defined the borders of all of the back yards in a neighborhood, many more fragments would be connected, increasing the ecological value of each by many times. If all of the neighborhoods in a community did the same thing, the existing habitat value would increase exponentially.

How could we possibly achieve that level of coordination among individual property owners in our present cultural environment? Consider this: not too long ago many houses had two pyramidal yews flanking the front door and an arborvitae on both front corners of the house. If we could practice that level of conformity for no apparent reason, maybe it is not impossible that we could do something similar in the name of environmental enhancement.

TRANSITION ZONES

In ecological terminology, the area where two different habitat types meet is called an ecotone. Here spotted geranium of the woodland may mingle with the early goldenrod of the meadow, while meadow-rue (*Thalictrum* species) occurs only in this unique transitional niche. Vines, including the native fox grape (*Vitis labrusca*), often occupy these areas, crawling up the trees and sealing off the woodland edge from light.

Most would consider this last feature, the growth of vines, a bad thing. It's true that the vines can kill the trees along the woodland edge, and they create an unsightly tangle visually separating the peripheral trees from the woodland. However, the vines are also providing a benefit for the plant community inside the woodland. Most woodland edges in our landscapes were created by people as they cleared areas for development, farm fields, or roads. This suddenly exposes a plant community that was formerly living in deep shade to increased levels of sunlight. The vines seal out that light and protect the newly exposed plants from what for them is a very negative change to the habitat.

That may be all well and good, but most of us do not want a tree-strangling tangle staring back at us from the edge of the woods. Consequently, a more aesthetically acceptable version, one that fulfills a similar function, may be in order. In addition to vines, certain trees and shrubs often emerge as a response to woodland edge creation. Many of these plants are clonal root expanders, such as staghorn sumac, gray dogwood, and quaking aspen (*Populus tremuloides*). Due to their dense colony structure they also serve as light

inhibitors. Evergreen trees such as eastern redcedar (*Juniperus virginiana*) and American holly (*Ilex opaca*) can also occupy this niche, offering a dense foliar curtain in spring when the deciduous edge species have yet to leaf out.

It is also important to note that the edge effect I have described is most pronounced on south- and west-facing edges, where the sun angle results in deeper light penetration into the woods. East- and particularly north-facing edges may not require light-inhibiting growth, as little light penetrates from these directions.

There are also other aspects to consider when designing the transition from woodland edge to meadow. In a typical human-designed situation, the border between wood and field is a straight line and most often abrupt: that is, the transition is from full-sized woodland trees to knee-high grasses and wildflowers or lawn in a matter of a few feet. Such an abrupt, rectilinear transition isn't visually graceful and it doesn't offer the greatest benefit to wildlife. A more natural transition includes shrubs at the woodland edges, shrubs

This meadow grades to shrubs and then a woodland beyond, providing a soft, gradual transition that is both aesthetically pleasing to humans and ecologically beneficial to wildlife.

In another area of the same meadow, the woodland and meadow wind sinuously together. Besides providing structurally varied types of cover for wildlife, these ecotones beg for exploration.

that along with pioneer tree species expand outward into the meadow to make a series of peninsulas of woody vegetation. A sinuous woodland edge of this kind creates a series of pockets of semi-enclosed grassy areas, which is ideal habitat for some species of birds and insects that won't inhabit a straight, abrupt transitional area. The sinuous edge is not only better for wildlife, it also has a more organic, dynamic appearance and it invites exploration, luring the visitor along to look into the pockets one after another. A more gradual transition, with shrubs intervening between the trees and the meadow, is also more graceful in appearance.

THE AESTHETICS OF THE LANDSCAPE PATTERN

In many areas throughout North America, it is the interplay of woodlands, open landscapes, and the transitional areas where they meet (edges or ecotones) that create the natural landscape patterns. You would do well to observe and mimic this in your own work, for a graceful and functional mix of these features can help determine whether your design will be aesthetically successful, even before you begin selecting plants. This can be especially fruitful for small properties, where manipulation of landscape patterns can create the illusion of more space. If you look out at a meadow, for instance, and at the far end it disappears behind a drift of shrubs, you can't tell where the meadow ends: in the imagination it can run on and on, far beyond its actual limit. Many of the natural patterns I described, including sinuous woodland edges, meandering shrub drifts that connect fragmented habitats, and the dynamic visual interplay between interlocking habitat arrangements, lend themselves quite well to this sort of treatment.

Habitat types arranged in a manner that manipulates views—both good and bad—also play important aesthetic and practical roles. For example, you can use tall tree or shrub compositions to frame and accentuate attractive views or to block undesirable ones. Your initial site analysis will have determined not only the location of unwanted

Site condition variations in this large seeded meadow enabled the creation of specialized drifts that contribute to the meadow's aesthetic appeal.

views, but also the minimum height of screen that can block them. I say "minimum height" because too often unnecessarily lofty screen plantings take out the good with the bad. While tall trees may indeed be needed to block views of large houses across the street, there is no reason to plant a row of spruce trees that will reach a height of 60 feet to screen a view of the street and passing cars, which may require a height of only 6 feet or less. This is particularly true if desirable views of off-property woodlands will be also removed from view. While removing the street from the visual picture, a relatively short native shrub drift can actually appear to be a woodland edge, visually connecting the property with the distant natural area and creating an optical illusion of sorts.

Choosing the landscape pattern the gardener will pursue also involves interweaving the practical with the aesthetic. Planning how the landscape will accommodate activities it is intended to support are of course part of the equation. The menu of activities varies with the individual and may include a play space for children, a shady retreat for summertime sitting, or perhaps emphasizing or creating views to enjoy from the house or a terrace. This is a part of any successful style of landscape design. In an ecologically based design, however, the gardener needs to be sensitive to the nature of the site.

DEVELOPING CRITERIA TO GUIDE PLANT SELECTION

Among the most fundamental decisions in an ecological design are the choices of which plants and which complexes of plants you introduce into each area of the landscape. While much of this is based on ecological analysis, taking your cues from nature does not mean you abdicate all decision-making.

Within the vegetation a site may support, there are generally many alternatives to choose from. Some of these alternatives may recommend themselves purely on the basis of practical needs and aesthetic preferences. How you select, fine tune, and integrate these choices with the site's existing ecological characteristics is what determines the ultimate look of the garden or landscape and the path down which it will evolve. Generally, I begin my decision-making about what I will plant and where by considering the plants' level of aggressiveness, nativity, aesthetic character, usefulness for stormwater management, and level of maintenance.

AGGRESSIVENESS

Some plants and complexes of plants are more aggressive than others—they are robust growers and often spread vigorously. Such plants should be paired with those of a similar level of competitiveness. Otherwise, the more aggressive members of your planting will overwhelm and push out the less aggressive species.

Keep in mind that you are likely to want to vary the level of plant aggressiveness from one area of your landscape to another, matching the plants to the situations. You may not want to include some aggressive sumac (*Rhus* species) or brackenfern (*Pteridium* species) in the more mannerly area in the front yard, but quite likely there will be a spot for them in the woodland edge planting you install in the back forty.

You may be planting an area that is heavily infested with weeds and invasive plants, with a lot of weed seeds in the seed bank and surrounded by weedy areas. If you don't have

Aggressive native plants like wild bergamot, oxeye sunflower, and ironweed (*Vernonia* species) are needed in soils with a large weed seed bank, like this one on former agricultural fields.

much time or resources to invest in weed control, then quite likely you will want to design highly aggressive plantings that can do battle successfully with their unwanted neighbors. Such bare-knuckle plants are commonly labeled disparagingly as "garden thugs," but if natives, they are invaluable not only for fighting invasive species but also for holding down more remote areas of a landscape with a minimum of maintenance.

If, however, you are aiming at subtler effects, an intricate garden to be viewed from close-up, and you wish to create a maximum of vegetative diversity, then you will probably want to plant less aggressive species that can coexist without overrunning each other. Often, I plan such areas for the immediate vicinity of the house, where the visibility is high and the spaces are more compact and so easier to give the higher level of maintenance that less aggressive plantings tend to require. In broader areas farther from the house, I often fill with more aggressive plantings that require less maintenance.

If you want your plantings to be robust and quick to establish but also exhibit a more gardenesque quality than the wild perimeter areas, then you may wish to design for moderately aggressive plantings. On small properties or where an overly wild appearance is not desired, this sort of transitional area includes much, if not most, of the landscape.

NATIVITY

Because we are attempting to work with and reinforce the natural habitats and plant communities of the place, the stock in trade for garden ecologists is native species. However, the seemingly simple term *native* is subject to widely differing interpretations, and you'll need to decide how you define it before you start planning your plantings.

By native do you mean native to your immediate locale? Native can also signify indigenous to the ecoregion or even the more general region, such as the Northeast, the Southwest, or even as broad an area as the West Coast. Any of these choices are acceptable, but remember that the plant's adaptability to a specific habitat type will most likely determine its ability to thrive and proliferate on your property. A plant that naturally flourishes in the rain forests of Washington State doesn't care if it is growing in Oregon, Washington, or British Colombia, as long as it is in its preferred habitat.

Finally, native can include anything indigenous to the North American continent. Frankly, this doesn't mean much from a survivability standpoint, as this continent has such a wide variety of habitats. A maple from the temperate forests of Japan is certainly better adapted to a garden in the northeastern United States than is a tree from the Arizona desert.

Taking North American nativity as your litmus test for acceptability does help to reduce the likelihood that you'll introduce any invasive exotic plants into your landscape, and it gives you the broadest latitude to experiment with diverse planting. By juxtaposing plants from different ecoregions, however, you greatly reduce the chances that your landscape will exhibit the highly integrated growth habits, growth periods, and chemical interactions contained in plant communities that have coevolved over millennia.

Ultimately, though, don't be too rigid. If you can't resist the white blossoms and delicate aspect of a Carolina silverbell (*Halesia carolina*) tree and want to install one in your yard in New Jersey, you shouldn't let the fact that this tree is native to the southern Appalachians

This paperbark maple, a native of Asia, was here before the herbaceous plantings. Because it was growing well, provided desired shade, and is by no means invasive, it was incorporated into what is otherwise a woodland garden of regionally native species.

and Piedmont dissuade you (Carolina silverbell is winter hardy to southern New England). If the beautiful autumn blooms of Japanese thimbleweed (*Anemone hupehensis*) will make you happy, then by all means find a home for it. There is no reason to exclude exotic species that have proven not to expand and dominate in the wild, such as paperbark maple (*Acer griseum*). Do keep in mind, though, that the farther your planting choices stray from home, the less benefit your landscape will provide to native pollinators and other local wildlife, particularly those specialists that have coevolved to use only specific native species.

In some locations, particularly urban areas, the ecological conditions that supported local plant communities may no longer be present. In this case, a gardener has three basic choices: restore the prior ecological conditions, work with the current disturbed conditions, or assemble a novel community of native and exotic plants.

Restore the prior ecological conditions

In this approach, you determine soil characteristics (including type, layer configuration, biota, and hydrology) that were present on the property prior to disturbance and then recreate them. Sometimes this is practical on small properties. Often, however, it is impractical, and in some cases it is ecologically problematic to truck in large amounts of soil and organic material.

Select plants that occur in the habitat most closely resembling the current disturbed conditions

If the topsoil layer has been lost to construction activities and you're left with a poorly drained clay subsoil, an ephemeral wetland plant community may serve as a model. This community can withstand the poor drainage from inundation during rainy periods and the dry conditions during drought. If you've been left with an impoverished gravel bank, a dry

To frame a lawn panel, a traditional landscape motif, native grasses and wildflowers are planted in distinctly formal patterns.

meadow community including little bluestem and butterflyweed will be willing participants in your new landscape. Even a soil consisting of mostly urban fill may have a counterpart in the stony scree landscapes that occur in some areas with shale-derived soils.

Abandon the native plant community model and fashion a novel community of native and exotic plants

Even when plant compositions are designed with species from a variety of locales, the characteristic niches of a typical plant community can be identified and filled, so that the resulting landscape contains some of the stability and resiliency of historically coevolved communities. These assemblages, which take the geographically mixed approach of traditional gardening to a more scientific level, are referred to as novel plant communities.

AESTHETIC CHARACTER

Just because garden ecologists base their designs on nature doesn't mean that they can ignore the sort of aesthetic decisions familiar to all gardeners, such as framing views or balance in compositions. The net result should be attractive as well as ecologically sound.

I remember attending a lecture by a landscape architect who was a native plant enthusiast. His house was located in a small New England village, and he had planted the green strip along the sidewalk with coarse, straggling poplar trees. The landscape architect ascribed his neighbors' dislike of this feature to ignorance. Based on the photographs, however, I suspected it had to do more with his bad taste, for this feature was simply ugly. The trees were native to the region, but out of character with the location.

Without attempting to cover all aspects of landscape design aesthetics, garden ecologists should consider the following aesthetic aspects: wildness versus formality, textural character of the plants, maintaining seasonal interest, and color preferences.

Wildness versus formality

How groomed do you want your landscape to appear? Quite likely, you'll prefer different levels of wildness or formality in different areas of the garden, perhaps a more formal look immediately around your house and a less groomed, more natural look as you move toward the outskirts. A formal layout, even a traditional look in some areas with plants in beds or borders, doesn't preclude the use of native plants or the benefits that they bring to wildlife. Such transitions can be gradual, the formal slowly grading into a wilder look, or they can be abrupt, a dramatic juxtaposition.

One useful design technique involves using native plants that have been tamed somewhat by plant breeders and so partake both of the wild and the cultivated. Such cultivars are often more compact than the wild species type, more floriferous, and sometimes disease-resistant as well. These choices can be very useful in making a transition from wild areas to more formal ones.

Textural character

Do you want the delicate look that wispy, fine-leaved grasses and meadow flowers can bring to your landscape? Or do you prefer the gutsier, more assertive look of bold-leaved plants? Alternatively, you might like the dramatic contrast that comes from juxtaposing the bold and the delicate.

Again, you will likely want to vary the textural character with the area of the garden or landscape. Bold foliage textures, for example, such as those of native magnolias (*Magnolia* species) or bottlebrush buckeye (*Aesculus parviflora*), can make a small space seem claustrophobic; in a situation of this kind, you'll probably want the airier texture of native trees such as serviceberry (*Amelanchier* species) or American hornbeam (*Carpinus caroliniana*).

For even more delicate effects, you can use meadow grasses such as little bluestem or prairie dropseed (*Sporobolus heterolepis*). One of the best features of such a planting is the wispy character of grasses moving with the breeze, and the addition of flowers can detract from this effect. However, flowers with narrow, grass-like foliage and smaller blossoms such as Virginia mountainmint or blazing star won't conflict with this delicate effect. Coarser leaved, larger flowered eastern purple coneflower (*Echinacea purpurea*) or oxeye sunflower (*Heliopsis helianthoides*) can be planted in isolated drifts to add a contrasting bolder effect. I've found that in a meadow, a balance of two-thirds airier plants with one-third bolder plants makes an attractive contrast.

Seasonal interest

Your planting choices should reflect whether you are designing a garden for a summer home or a place you enjoy throughout the year. Plant lots of summer bloomers, for instance, in a summertime garden, and in the winter garden add plants with evergreen foliage or intriguing structural form and colorful bark and twigs.

Once again, you may apply this criterion differently in different areas of the same landscape. For instance, include more winter-showy plants in the view from the breakfast room, where you'll have your cup of coffee on winter mornings, and

more summer bloomers further from the house, where you like to stroll or sit during that season.

Designers tend to think only in terms of evergreen for winter effects, with a few brightly colored stems thrown in for good measure. Although good effects can be achieved in this way, in our natural landscape the bulk of the colors come from the reds and oranges of grasses, the blacks and beiges of seed heads, and the blacks and grays of tree trunks,

The sparkling orange hues of a little bluestem meadow are stunning in the slanting winter light and provide a dynamic contrast with stonework and nearby evergreens.

with evergreens just scattered here and there. Replicating these colors and patterns in gardens can create beautiful winter vistas. I should add that not all evergreens are good choices for winter interest. Rhododendron leaves curl and droop in response to arctic blasts, and nothing makes me feel colder on a frigid morning than looking out the window at a bed of cringing pachysandra foliage.

Color preferences

Obviously, if you hate red, you should avoid planting red-foliaged and red-flowered gardens. It's *your* landscape. One caution I would add, though, is if you are planning an open meadow, you should come to terms with yellow, because there are many, many good meadow plants that have yellow flowers. It doesn't have to be the dominant color of the meadow, but if you exclude yellow entirely, you'll find it much more difficult to create a well-balanced, integrated plant community.

It's easy, by the way, to focus too much on flower color. The structure and architecture of plants is just as important to the look of a composition; a tangle of mismatched perennials struggling to elbow each other out of the way won't be attractive no matter how carefully the colors are blended. I do have a confession to make here. On most of my projects, considering color combinations is the last intellectual activity on my design to-do list, if it is there at all. Have you ever wandered into a lush woodland bursting with native spring wildflowers and gasped, "What a hideous color clash?" Enough said.

STORMWATER MANAGEMENT

Conventional landscape design too often focuses on draining stormwater off the property, which tends to create flooding and stream bank erosion downhill in the watershed. In a healthy system, the water deposited by precipitation infiltrates down through the soil, replenishing crucial underground aquifers and preserving water resources. Ecological garden design partially remedies this by replacing lawn, which retains relatively little of the water that falls on it, with other plantings that are superior at absorbing water. Dips and depressions, instead of being regarded as a problem to be eliminated, are turned into rain gardens that further promote infiltration.

Rain gardens can be simple planted depressions or contain highly engineered associated structures. We won't go into the latter here, as there are excellent resources for further investigation. Species that adapt best to rain gardens are those that are native to ephemeral wetlands, areas that are wet only part of the year and dry the rest.

LEVEL OF MAINTENANCE

A consideration that should be fundamental to your design is the level of maintenance that the landscape will receive. Ecological design, when properly carried out, greatly reduces the need for maintenance but does not eliminate it altogether. Clearly, you don't want to create a garden whose maintenance will demand more labor than you want to contribute yourself or can afford to pay for.

Coming to a realistic assessment of what degree of maintenance you can furnish will affect every feature of the design. If, for example, you know that you will have little time to devote to maintenance, you should design with more aggressive, self-sufficient plants (knowing that the downside of this will be a less diverse plant community). Or you may plan to install the landscape in phases. A newly installed landscape typically requires more intensive maintenance because the disturbance brings on a flush of weeds that doesn't taper off until the plantings have knit together; in short, it mimics the pattern of initial floristic composition found in the natural response to ecosystem disturbance. Rather than trying to install the whole landscape at the same time and thus facing weeds everywhere, it makes sense to divide the design into sections, completing one and waiting for that to establish before you address the next. In general, plantings with a robust level of aggressiveness, if sited properly, will be your best tool to reduce maintenance.

Limiting maintenance also requires foresight: you should formulate the design to suit planned management procedures. For example, you may plan to cut over or mow some area at a certain height and at a certain time of year to prevent seed formation of a particular weed. This plan should be reflected in the design, which should exclude from the ground layer any plants that exceed this mowing height, so that you can carry out the weed-inhibiting cutting without damaging new plantings.

Coordinating design and management can be a very powerful tool in ecological garden design, allowing much larger areas to be effectively managed than would likely be possible through traditional practice in which the design and management of the landscape are disconnected.

CONSIDERING CIRCULATION

Minimizing turf areas in ecological garden design has an important impact on the design of pathways. Where lawn dominates, garden users can stroll off in any direction they desire. However, when lawn is replaced by meadow with its taller vegetation, pedestrians are confined to paths. Consequently you now must consider how the paths within the meadow will be placed to allow for direct access to all areas of the property. In contrast, when tangles of invasive vines and other weeds are removed from a woodland, access is increased, and paths become important as a means of leading users through the landscape in a practical and comfortable manner, helping them to avoid wet areas and navigate steep slopes, while also providing access to streams and any other water elements that the landscape may offer.

This path, inviting exploration, leads from an open meadow into a birch grove.

Beyond the practical, the experience of traveling through and lingering in the landscape can be greatly enhanced by carefully thinking through the placement of paths and stopping points. Paths should take users through all of the varied habitats—large and small and natural or created—that are present on the property. Paths can also lead to particularly inspiring physical elements, including old growth trees, rock outcrops, and sculptural tree snags. The list can go on, of course. These elements are essentially free, a bonus that needs only to be accessed and appreciated.

Locate paths and stopping points to provide access not only to a diversity of plant communities but also to a diversity of experiences. A path can wind through a dense shrub thicket and open suddenly to an expansive meadow view. A created space within a thicket of tall grasses like big bluestem can provide an intimate view of the multicolored stems and the soothing sounds of the grasses rustling in the wind.

PLANNING FOR SUCCESSION

It's important to understand that any design you develop must have flexibility built into it. That's because the indigenous landscape is a constantly changing system composed of plants, animals, microorganisms, and soils. Plants are not isolated entities, but participants in a system constantly in flux. If you visit a beautiful spot in the woods and inventory all the plants there, when you return in ten years some species will have dropped out and some new ones will have come in. In addition, different types of systems change at different rates. The annual meadow immediately resulting from a disturbance may last for only one year (unless the gardener intervenes), whereas an unmanaged perennial meadow may last for ten years before yielding to pioneer forest species. In contrast, an old oak and hickory forest may last for hundreds of years if left undisturbed.

Once these changing systems are understood, designers can decide which aspects to encourage, discourage, or manipulate to fit the requirements of the site. Understand that whatever plan you develop, it isn't a blueprint. That sort of controlling design, in which every detail is specified and fixed, the staple of traditional horticulture, creates a static photograph frozen in time, a landscape forever doing battle with the forces of nature in which change is the constant. Ecological landscape design, in contrast, is the beginning of a dynamic process and a partnership with nature.

In most parts of North America, meadow, shrubland, and woods are the three primary stages of succession. Each of these designations is basically describing a vegetative end point, which when reached will be arrested through applied management. What are the starting points of these stages? A woodland canopy can be created relatively quickly by planting large or fast-growing trees densely throughout a field. In this case, both the starting and end points are woods. Alternatively you may plant the field with scattered small trees and shrubs and allow woods to develop over time. In this case the starting point was the oldfield stage between meadow and shrubland, while the end point is woods. Or you may simply plant a meadow and allow the shrubs and trees to naturally recruit, which will happen in much of North America. Specifying the beginning and end of the succession process for each area is the last element to determine during the early stage of the design process.

SYNTHESIZING THE INFORMATION INTO AN INITIAL PLAN

Ecological landscape design is a multilayered process, a balance of many kinds of input. It can be helpful to think of this as a series of overlays: starting with the raw ecology of the site, you overlay your own tastes, the practicalities of maintenance and management, the needs of wildlife, and the ways in which it all interacts to create an initial plan. In some areas this may be highly designed, with the actual arrangement of plants specified, and in others nothing more than planting lists. Understand, though, that whatever plans you create, the site itself is going to take a hand in adapting and transforming it as you go along. Change is the constant.

right Due to the presence of crownvetch (*Securigera varia*), a problem weed with extended seed viability, portions of the meadow were designed as grass only, so that broadleaf-specific herbicides could be used to control the vetch.

facing page The conceptual design of this site features a variety of habitat types from upland to riparian woods and meadows. Analyzing the various physical characteristics across the site allowed plants to be matched to their appropriate habitats.

SCALE: 1"=20'-0"

0   20   40   60

N

CRANMER RESIDENCE
1650 NORTH RIDLEY CREEK RD.
MEDIA, PA. 19041
NATURAL AREAS PLAN- UPPER PROPERTY

PAGE
1
OF 1

top Excluding broadleaf flowers allowed the crownvetch to be controlled and native grasses to be established.

bottom Late-stage meadow species, including mountainmint, wild bergamot, and Indiangrass, thrive in the meadow.

top The plan called for an area along the driveway to succeed to tree and shrub groves, which diversified existing habitat while helping to frame views and create a more layered experience.

bottom Bald cypress was planted in a wet pocket at the woodland edge.

INSPIRATION FROM UNCLE MAX

When I visited my Uncle Max in upstate New York, we did not talk about gardening or ecology, yet a story he told me has deeply affected how I view my work. Max had a varied and accomplished career in the world of music. As a young assembly-line arranger on Tin Pan Alley, he brushed elbows with Depression-era songwriting legends like George Gershwin. He repaired instruments for virtuoso string players from the New York Philharmonic, and worked as the favored music copyist for Leonard Bernstein. He even recounted a phone conversation with Igor Stravinsky, during which Max asked the composer about his preference for notating a particular passage in one of his scores. "It doesn't matter," Stravinsky stated matter-of-factly. "The musicians can do whatever they want there." As a composer myself, I ate these stories up. I did not expect, however, that one would relate so directly to my experiences in the landscape world.

Max was also an accomplished violist, and he recounted an experience he had while rehearsing a Haydn string quartet with his ensemble. This was just a practice session, not a performance, so the musicians were mainly focused on the mechanics of their parts. Yet, at one point, the music became so beautiful and profound the entire quartet simultaneously stopped playing, awestruck. I think that the emotional impact of that moment came not only from the music itself, but also from the connection the musicians made through the act of playing it. The sound that floated across that room was not all them, nor was it all Haydn. It was the collaboration among the musicians and composer that affected them so deeply.

Max's story made me realize that I had often experienced similar moments while working in the landscape. These moments generally didn't arise from an entirely composed garden element, but from a spontaneous event that occurred only in part as a result of my efforts, like when a beautiful and rare purple-flowering fringed gentian (*Gentianopsis crinita*) appeared in a meadow planted a decade earlier. Or when a stately 7-foot-tall cup plant wandered from my clients' planted meadow into their pool garden, placing itself perfectly in the center as if it was the intended focal point. Or when I intentionally burned another meadow that contained a few sporadic spiderwort (*Tradescantia* species) plants, thus igniting a 7-acre sea of blue spiderwort blooms the following year. None of these occurrences were all me, but neither were they all nature. Like Max's ensemble playing Haydn, they were the result of collaboration.

Obviously occurrences like these are much less likely in traditional landscapes, where plant selection and arrangement is the exclusive domain of the designer and natural recruitment plays a small or nonexistent role. While reflecting on my landscape experiences it occurred to me—surprising as it may seem—that connections with nature were also less intense when visiting spectacular wild places like Yosemite National Park. Of course, I was enthralled with the dramatic scenery when I visited the park, but I had no personal connection to this landscape as I had played no active role in its development. I realized that working in direct concert with natural processes, as I had frequently done in the places where I live and work, had fundamentally changed the way I related to these landscapes. The

This rare purple-flowering greater fringed gentian showed up in a 10-year-old planted meadow. Such surprises are what make ecological gardening so stimulating.

connections there were more intimate and intense than those a fully planned garden or an entirely wild natural area could provide.

Shortly after my story-laden visit with Uncle Max, I had an encounter with a landscape architect who had brought me in to create a meadow on one of his projects. The meadow was to be quite large, and I designed many different seed mixes for various microhabitats on the site. The landscape architect decided to create a series of seasonal color charts for each of these mixes. This extensive document contained many pages of beautifully designed charts showing the color combinations that each mix would produce at each time of the year.

He presented it to me, saying, "I just wanted to get some insight into the thought process behind your design." Actually, my thought process had largely consisted of selecting the plants that would live on this weedy and difficult site. Formulating a color scheme had played virtually no role. The landscape architect's question reminded me of the question Max had asked Stravinsky about notating the musical passage, except my answer was "It doesn't matter. The plants can do whatever they want there."

above The work of this bumblebee is integral to the continued presence of greater fringed gentian in the meadow ecosystem.

right This meadow composition reflects the interplay among the site conditions, the various plants' characteristics, and the work of the garden artist.

DEVELOPING A SYNERGISTIC PLANT LIST

You have worked out the overall design for your property and set some general plant selection parameters. Now it's time to get specific and start selecting the exact plants you will fit into this framework.

Golden ragwort spreads reliably by rhizomes to form a dense, weed-suppressive groundcover.

SELECT THE PLANT COMMUNITIES

This begins with thinking about the plant communities that will inhabit each ecological zone. A woodland zone, for example, which may consist of existing woods and an area you intend to reforest as well, could include several different habitat types: a wetland, a dry ledge, and maybe some upland woods. You need to identify these different areas within the zone and then determine which plant community is best adapted to each of those areas. The same process applies to shrubland zones, hedgerows, and meadows, as each may include areas with different growing conditions and so require a different plant community.

The habitat and natural features base map that resulted from your site analysis efforts should guide you in this regard, as it will show the locations of the physical habitats that exist on the property. Effectively matching the best adapted plant community or communities with each of these physical habitats will be a critical activity in the planning process.

The first step is to determine which ecoregion you are in. This information is relatively easy to obtain online from the U.S. Environmental Protection Agency (epa.gov/wed/pages/ecoregions.htm).

Next you will play matchmaker and determine which plant communities within your ecoregion are best adapted to the physical habitats that occur on the property. You can find guidance in several resources, such as the *Landscape Restoration Handbook* by Donald Harker and colleagues, which provides maps of the entire United States with ecoregions clearly shown. It also names the plant communities with component species lists that correspond approximately to the Environmental Protection Agency's ecoregion maps.

Local floras, available in print or online, will generally show the most detailed information on native species and where they grow within an ecoregion. Those that group plants by community, as opposed to by family and genus, will be the most directly useful.

Flora dot maps can also be found online or in printed editions. These generally consist of a state or provincial map for each species, with dots showing the recorded locations of that species. These will tell you whether the plant is native to your specific locale within the region. Flora dot maps can also tell you whether that species is common or rare in your area based on the density of dots.

Finally, the observation of natural areas is perhaps your most effective educational avenue. Find out where natural areas exist that have similar physical habitats to your property, and inventory the vegetation. This can be done on your own with a field guide; or, better yet, take a nature walk with an ecologist, who will likely have much of the information you are seeking in his or her head.

FINE TUNE THE SELECTED PLANT COMMUNITIES

Once you've selected the plant communities you want to use in your landscape, the process of fine tuning begins. Let's say you've decided to include an area of meadow. You analyzed the sun, soil, topography, and drainage and determined that a particular type of upland grassland community will flourish in these conditions.

That doesn't mean that you have to include every species that naturally occurs in such a community in your planting. You may decide that some don't fit the criteria you established in the course of your design process. Some species, for example, may be too aggressive and by dominating will reduce the level of plant diversity you are hoping to achieve. Wingstem (*Verbesina alternifolia*), for instance, is a bona fide member of some eastern grassland communities, but it's far too aggressive for most gardeners to want to include, unless the object is low-maintenance suppression of invasive plants in a less intensively managed area. Other members of your selected plant community may be insufficiently aggressive, unable to cope with the weed pressure you know that your garden will experience, or of a color you don't like, too wild looking, or too delicate in appearance.

You may also wish to include some plants that don't naturally occur within the community you've chosen. Early in the design process you will have established how strict your interpretation of native will be for each area in the landscape. Whether you are including species that are found within an ecoregion but not in your model plant community or are incorporating plants from other ecoregions, you should investigate how others have fared using the plant in a wild or semi-wild setting. If this is not possible, experiment with a few individuals and add more later if the results are positive.

But what of using exotic species from other parts of the globe? Maybe you're nostalgic for the sweet-scented lilies or lilacs your grandmother grew in her garden and you want to include those. Or you want to boost the blues or even include some edible plants to

supplement your groceries. As long as they are not on any invasive species lists, there is no reason to exclude them. However, I usually associate these plants with more highly cultivated areas, such as around the house, and preserve the wilder hinterlands for the species that grow there naturally.

All the major spatial niches from canopy level to subcanopy level to the ground layer are occupied in this woodland edge planting, where herbaceous plants were added following removal of invasive species.

A major advantage of starting your planting list with a native plant community is that the species within such a grouping tend to act synergistically. Because they evolved together over many generations, these plants not only coexist comfortably, they commonly benefit each other. When you plant a natural community, you aren't juxtaposing a miscellaneous group of competitors, as you might in a traditional garden full of plants from disparate regions. Such disparate plants haven't had a chance to coevolve and form relationships. Instead, you're planting a team.

Selecting species for your planting list involves considering several criteria. I find the following items useful to ensure that I don't overlook any important points.

FILL THE ECOLOGICAL NICHES IN SPACE AND TIME

I've already alluded to the way in which different plants within a community specialize to fill different physical niches with their belowground and aboveground growth. In fact, a defining characteristic of any stable plant community is that all such niches are filled. You must make sure to do the same in the landscape you are creating. A basic law of almost any ecosystem is that if nothing is currently growing in a given space, something soon will. The more spaces you leave unoccupied, the more opportunity there is for weeds to enter: essentially, you are guaranteeing yourself a high level of maintenance. Fill the niches, and you forestall this problem.

Filling all the spatial niches requires three-dimensional planning, for plants grow next to each other, above each other, and below each other. Even a 3-foot-tall meadow has a multilayered structure. This is also evident belowground, where fibrous rooted plants occupy the soil surface and coexist with deep tap-rooted plants holding down the fort below.

This matter of filling niches provides an obvious contrast between an ecological garden and a conventional one. Examine a healthy native prairie and you may find five, six, seven, or eight or more plants per square foot. However, look in a conventional horticulture guide and you'll find recommendations to plant perennials 12 or even 18 inches apart. This difference originates in the character of the plant selection: the garden ecologist is planting a coevolved community, whereas the conventional horticulturist is typically using plants that just met, that haven't had a chance to develop cooperative relationships.

In addition to filling the various spatial niches, you need to fill the niches in time. First is seasonal time. Some plants are active in warm weather, whereas others are most vigorous during the cool seasons, particularly spring. Whatever type of planting you are planning, remember that including both early- and late-season plants ensures that there will be no seasonal window of opportunity for weed invasion.

A second type of temporal niche, the successional niche, is measured in years. Annuals, biennials, and fast-growing perennials establish cover during the first year or first few years, whereas long-lived plants may not have a serious visual presence for many years.

Although the vegetative layers in a meadow are less visible than a woodland's architecturally distinct layers, spatial niches still exist in the meadow, from the tallest growth at the canopy layer to prostrate groundcovers. When spatial niches are filled, it is difficult for weeds to become established.

With woody plants, the schedule is stretched, as virtually all require several years to become established. Nevertheless, there are relatively fast growing trees, early successional pioneer species such as birches and poplars that surge to maturity within a decade but age and die just as quickly. Maples and oaks, for example, are slower growing, but more enduring late-stage forest canopy species. In any area you are trying to return from open landscape to woodland, you should include both hare and tortoise species.

Because even the fastest growing woody plants take a few years to establish themselves, blanketing the area you are reforesting with weed-resistant groundcovers will require resorting to herbaceous plants. Typically a new planting of saplings won't provide enough shade to foster genuine woodland groundcovers. Instead, the initial groundcover should be a meadow composition. A dense seeded meadow planting can be a cost-effective and aesthetically satisfying way to stabilize the site in the interim, while it is still sunny. As the trees mature and shade increases, the meadow will begin to phase out and can be replaced with ferns and woodland wildflowers, which may also naturally recruit into the area.

In any case, whether the vegetation you are planning is woody or herbaceous, woodland, shrubland, or meadow, you will need fast-growing plants to cover and occupy the

This photograph captures a moment in time. The yellow blackeyed Susan, a biennial, and the white-flowered eastern daisy fleabane, an annual, are sun-loving meadow species. As the trees mature and provide more shade, these wildflowers will be replaced by more shade-tolerant herbaceous species.

A matrix of nitrogen-fixing blue wild indigo and white wild indigo create a dramatic scene in early summer. Belowground, nodules on the indigos' roots produce nitrogen that will become available to surrounding plants.

soil before invasive species move in, and long-lived, slow-growing plants contribute to an enduring mosaic of vegetation. I find it helps to think of this process as a sort of ecological relay race, with short-lived species passing the baton to long-lived ones in a calculated progression.

INCLUDE NITROGEN FIXERS

Members of the bean family (Leguminosae) play an essential role in many plant communities. Working in concert with bacteria that live in nodules on their roots, nitrogen fixers have the ability to take nitrogen from the atmosphere and convert it into a form that can be used by plants. Nitrogen is the nutrient that plants use in the greatest quantity. The importance of this ecological function is apparent in the number and diversity of this family, which includes some 18,000 species ranging in size from tiny annuals to large trees.

Horticulturists usually provide plants with adequate nitrogen by adding fertilizer. The amounts and rates of these applications are typically based on guesswork. In a traditional garden, which juxtaposes plants from varied continents, ecoregions, and habitats, it's

almost a given that some species receive too much and others too little. Nitrogen-fixing legumes, in contrast, have coevolved with the other members of their plant communities. If the appropriate species are intermingled throughout a community, the nitrogen brought into the soil environment by the legumes will be the optimal amount for the overall health of the system.

Woe to the weed that would like to become established in this designed meadow. The density of growth provides few, if any, opportunities for weeds.

CREATE A DENSE GROUND LAYER

Dealing with weeds is by far the most time-consuming aspect of landscape maintenance. But if you smother them with well-adapted, vigorous groundcovers, there will be no need for fabrics, grub hoes, and mulches to fight off weed invasion. Broad seas of mulch, a too common feature of contemporary North American landscapes, are an invitation to trouble. Covering a bed with 1 or 2 inches of shredded bark or pine straw will deter the growth there of less aggressive species, but it will not long deter neighborhood bullies such as Canada thistle, Japanese knotweed (*Fallopia japonica*), and Asiatic tearthumb (*Polygonum perfoliatum*). And once they gain a foothold in the mulch, the bullies will be hard to displace. A multilayered, densely planted groundcover layer, composed of plants with complementary aboveground and belowground growth habits, will be far more successful at inhibiting weed invasion than any mulch.

The next time you weed your garden, note what percentage of your time was spent in open areas and gaps in the plantings. You'll likely find that you spent very little time weeding where plants were well established and covering the ground. That is because seeds of most species require soil that is exposed to sunlight to germinate. Minimizing these gaps deprives these plants of this condition and is by far the most effective thing you can do to reduce the maintenance in your garden.

If you also design this ground layer for succession of bloom and contrasting foliage texture, you create a reduced-maintenance landscape that suggests the diverse tapestry of our native groundcovers while achieving an artistic and colorful composition.

INCLUDE PLANTS WITH MULTIPLE REPRODUCTIVE STRATEGIES

A traditional gardener's definition of success is when a plant survives, but an ecologist doesn't view that plant as a success unless it also reproduces. What this means is that, as you assemble your plant list and design the arrangement of those plants in the landscape, you must plan not only for a first generation, but also lay the groundwork for a second and third generation and beyond.

This is in part a matter of necessity. When you reject the concept of beds, of restricting gardening to small discrete areas, and take on the landscape as a whole, then you almost certainly exceed your personal planting capacity. You simply do not have the time or the money to plant your entire ¼- or ½-acre yard. You need nature's assistance. In addition, planning for plants' self-proliferation also helps to ensure that your landscape is not on a one-way march to senescence, with no offspring there to replace those plants that inevitably drop out.

Certainly the growth of volunteers, chance seedlings, or offsets isn't unheard of in the traditional garden, but it's a relatively rare occurrence and an accident. In the ecological garden, such naturally recruited plants play a crucial role in the landscape's development. Self-propagation under those circumstances is much too important to leave to chance. It has to be addressed during the planning process.

Planning for self-propagation requires selecting your plants with a greater degree of fidelity to the habitat. That is, if you want plants to prosper and self-propagate, you must make sure that they are well adapted to the habitat in which you place them. If they're out of their habitat in a garden, you may be able keep them alive, but they are far less likely to self-proliferate.

Ecology-minded gardeners often refer to the benefits of including a wide diversity of native plant species. This is certainly true, but including plants with a wide array of proliferation strategies can be equally important when self-perpetuating landscapes are the goal. Plants have different strategies for propagation, and you need to include as many of those strategies into your landscape as possible. Understanding how plants propagate also allows you to maximize the speed with which they colonize. There are two basic reproductive categories: sexual reproduction, which is achieved by seeds dispersed using either the plant's own means or through external means, and asexual reproduction, which occurs through the expansion of roots or stems.

Seed dispersal by gravity

Plants such as spotted geranium propagate by seed dispersed by gravity, known to specialists as barochory. This means that seedlings typically appear right around the base of the parent plant, and they consequently expand into the surrounding landscape very slowly. You can increase the speed with which these species colonize by planting them in small, scattered patches throughout the landscape. The same number of individuals planted in one large patch would take much longer, as there would be little proximity to large portions of the area, illustrating the importance of arrangement on colonization rate. Another form related to gravity dispersal is ballochory, the forceful ejection of seeds by dehiscence and squeezing, as with the seeds of native American witchhazel (*Hamamelis virginiana*).

Seed dispersal by wind

Plants such as early meadow-rue have seeds dispersed by the wind. In contrast to the gravity-dispersed geranium, a few individual meadow-rues planted at the side of the garden that faces into the prevailing wind ensure that the whole garden will be seeded. This makes such plants ideal for gardeners on a tight budget.

Seed dispersal by small mammals and birds

Oaks produce acorns, which are often planted by squirrels and chipmunks. These small mammals generally bury acorns and other nut-like seeds no more than 30 yards from the mother tree. Understanding this dynamic can guide how you space nut-producing tree species, particularly if colonization by seed dispersal is a goal.

Understanding how a species disburses its seeds can also indicate whether you can expect that species to naturally recruit on your property without planting. If oak or hickory trees are present, there is a good chance you will receive free plants, courtesy of a squirrel. If not, you will have to plant them.

Berry-producing plants are generally propagated by birds. Of course, there is a time lag between when the bird eats and defecates. If the berries are the only bird attractant in your garden, there's a good chance the seed contained in the fruit will be deposited in droppings on someone else's property. But if you have included in your garden all of the bird's habitat features, such as protected perch and roosting spots (including snags) and a water source, chances are much better that the birds will fully digest and pass the seeds before leaving.

Rhizomes

Rhizomatous plants spread by creeping roots that continuously spawn new plants at their tips as they expand. Such plants, including Pennsylvania sedge and golden ragwort (*Packera aurea*), tend over time to form dense colonies. Rhizomatous plants expand into the landscape bit by bit, and like spotted geranium they should be scattered in patches

following pages Golden ragwort here served as the primary colonizer of a woodland ground layer where invasive vegetation had been removed. Species with less vigorous reproductive means would have been insufficient to compete with invasive species regrowth.

throughout the landscape to maximize the speed with which they fully colonize the area. If planted at close intervals in a hospitable spot, over time they will produce an uninterrupted groundcover. As a result you can plant them in a monoculture and still expect them to be weed suppressive.

One thing to keep in mind is that most plants tend to focus on a single strategy for proliferation. If a species readily produces seedlings, it is unlikely to spread by stolons or rhizomes as well. Conversely, plants that employ asexual reproduction strategies often do not produce much viable seed. Consequently, don't be fooled into thinking that you can seed a Pennsylvania sedge lawn. I wish you could. It is a wonderful lawn alternative, but unfortunately planting live plants, an obviously more expensive option, is the only viable method for establishing this particular species.

The stolons of wild strawberry, here with its diminutive white flowers, have wound their way through this meadow, rooting wherever they find a gap and effectively plugging the openings that weeds could use to gain a foothold.

Stolons

Stoloniferous plants such as wild strawberry (*Fragaria virginiana*) reach out with trailing stems through the surrounding vegetation. Once the stem encounters a gap, it roots into the ground, forms a dense basal clump of leaves, and then a new stem heads off to seek the next available gap, where the process is repeated. In effect this plant is seeking out the vegetative gaps, the spots where most of the weeds occur, and plugging them.

Horticulturists often consider stoloniferous species in their wild form to be bad groundcovers. Leaves on the stems of these plants are often spaced widely along the stems, making them appear sparse. This has often been remedied by creating cultivars that grow more densely, such as the green and gold *Chrysogonum virginianum* cultivar 'Allen Bush.' If their hope for the wild forms of stoloniferous plants is that they will form a dense groundcover, then the horticulturists are right, these plants are unsatisfactory and the cultivars are better choices. Because they don't devote their energy to sprouting dense leaves all along their stems, however, the wild forms can travel much longer distances. Consequently, if planted in intermingled species compositions, the straight species will form part of a lush, more complex groundcover while still preserving the gap-plugging ecological function that is characteristic of stoloniferous plants.

ACCOMMODATE WILDLIFE

A plant community does not function in isolation: it is the interactions with wildlife that turn it into an ecosystem. Birds distribute seeds from berry-producing plants, and small mammals carry and disburse seeds that stick to their fur throughout the landscape. Many people plant butterfly gardens so they can enjoy their presence and help the little creatures. But in the self-proliferating garden, pollinators and seed dispersers are crucial for seed production and ultimately the creation of additional plants.

In your selection of plants, you need to set the stage for these interactions. Much attention is given to providing food for wildlife, particularly birds and butterflies, and a great deal of information exists from numerous sources. But there are other important aspects to consider. Make sure to provide pollen- and nectar-bearing plants that will attract and nourish pollinators. Shrubs provide cover for birds; dead snags should be prized as nesting places for a variety of animals, such as woodpeckers and small mammals. Water, of course, is always a necessity and is most useful to wildlife when surrounded by dense cover that provides quick escape from surprise attacks by predators.

ARRANGE PLANTS SYNERGISTICALLY

Assembling the planting list is like putting together a guest list for a dinner party: we must figure out who should sit next to whom to create the best experience for all concerned. The key to arranging plants into harmonious and mutually beneficial compositions is to study how nature does this. In a relatively intact habitat with a functional plant community, individual plants compete but they rarely crowd each other. Instead, they tend to occupy complementary spaces.

Plants compete for different kinds of resources, such as moisture and nutrients, and the various kinds of competition all shape how the plants arrange themselves within the landscape. The biggest single source of competition for plants, at least in temperate, moister regions, is the competition for light. Sunlight is the primary fuel of plant growth, and plants have developed different strategies for securing an adequate supply. Some grow tall, reaching up above their fellows. Others have evolved broad leaves that are superior solar collectors, so that these plants can grow successfully in the filtered, less intense sunlight found beneath the taller plants. Some leaf out early to get the jump on competitors. This is a prominent pattern in woodlands, where wildflowers such as Virginia springbeauty (*Claytonia virginica*) or bloodroot emerge in early spring, bloom, and retreat into dormancy again before the trees above them are fully leafed out. It also occurs in meadows, however, with precocious though less vigorous species such as shootingstar (*Dodecatheon meadia*) growing, blooming, and going dormant before the warm-season grasses overshadow them.

In general, plants tend to segregate themselves into strata according to the intensity of sunlight they require. This is most obvious in woodlands, where you will find tall canopy-level trees such as maples, oaks, and pines; shorter understory trees including dogwoods and woodland shrubs such as viburnums (*Viburnum* species) that flourish in filtered light; and the low, shade-tolerant plants of the woodland floor such as ferns and wildflowers.

This hierarchy is replicated, though less obviously, in other types of plant communities. In meadows, for example, creeping plants such as wild strawberries and cinquefoils

Bees, wasps, butterflies, and other pollinators love milkweed plants because they produce ample nectar and even replenish it overnight, thereby supplying nocturnal and early-morning visitors. Milkweed species such as butterflyweed are also essential to the monarch butterfly, whose larvae are dependent on milkweed foliage for forage.

PLANT COMMUNITY SYNERGY

As soon as I began thinking of a garden as an eco-system, as a complex of plants and animals interacting with each other and their immediate environment, I found the old way of choosing plants to be inadequate. Of course, aesthetics are important in any garden—the result of your work must be pleasing to the eye—but just because the flower colors of two plants harmonize it doesn't mean the two organisms were biologically compatible.

The relationships among plant species that evolved together in the same habitat type can extend beyond lack of aggression. In many cases, such natural associates interact in mutually beneficial ways. A colonizing perennial such as Allegheny spurge (*Pachysandra procumbens*) isn't a challenger to the oak trees under which it is commonly found in the wild. Instead, the oak canopy provides the sort of filtered light that the spurge requires, and the spurge, with its network of stoloniferous stems, prevents soil erosion and traps leaf litter from the trees, creating a cool, humus-rich forest floor that is ideal for the oaks' roots. In many ecosystems, mycorrhizal fungi in the soil tap into plant roots, drawing sugars from the plants but increasing the plant's access to water and minerals, so that both organisms benefit. In some cases, mycorrhizal fungi actually link one plant with another, creating a sort of super-organism.

Sometimes the way in which different species support each other is more mundane, but no less valuable to the gardener. Culver's root, a summer-blooming meadow and prairie perennial native throughout the eastern half of North America, is popular among gardeners for its branching spires of white, pink, or blue flowers borne atop soaring stems that can reach a height of 6 feet or more. In a traditional garden border, these stems often must be staked or they will flop. In an ecologically designed meadow, however, they usually find sufficient support from their neighbors to remain upright.

above When amid grasses and other wildflowers that support it, Culver's root stands upright without requiring any staking.

above Without its grassland associates and in the rich soils of a garden setting, Culver's root requires staking.

(*Potentilla* species) cover the meadow floor, and bushy plants such as wild indigos make branching mounds maybe 2 to 3 feet tall. The meadow equivalents of canopy plants, tall native grasses such as big bluestem or taller wildflowers such as Culver's root, reach up to spread about and overshadow the plants below. All of these plants may be rooted together into a small space, but because they are growing at different strata, they don't crowd each other.

Traditional horticulture assigns an ideal spacing to each type of plant, so that the plants can grow up without impinging on each other. This is based on the idea of a garden as a collection of disparate individuals. A garden ecologist seeks rather to create a fabric of vegetation over the landscape, and different plants, if they cooperate like canopy and understory trees rather than compete, may be planted much closer—almost in the same hole at times—than is recommended in traditional horticulture.

One exception to this intermingled arrangement is rhizomatous species. They often form monocultures due to the highly competitive, exclusionary nature of their mat-forming root systems. If planted in a way that essentially replicates an observed natural pattern, rhizomatous shrubs like the native steeplebush spirea, white meadowsweet, sweet fern, or numerous sumac species can provide some of the easiest landscapes to manage once a closed cover has been established.

Filling all the niches helps to exclude weeds, and so does arranging the plants synergistically. In a woodland, shrubland, or meadow where all the strata have been filled, there are so many layers of plants an invader would have to penetrate that the planting, once mature, is notably weed-resistant. Planting synergistically provides aesthetic benefits as well: the orderly, if sometimes irregular, structure of a skillfully planted woodland, shrubland, or meadow presents a harmonious, graceful picture.

My clients prefer a natural landscape, so we planted trees to screen out the next-door neighbor's massive lawn.

opposite page When a neighbor stopped mowing fields adjoining a designed meadow in the summer, warm-season meadows grasses, like Indiangrass with its distinctive tawny seed heads, burst forth. As the planted trees mature and produce seeds, they too will proliferate across the landscape. And so the adventure will continue.

above Indiangrass and flowers create a subtle tapestry in my client's meadow.

pages 166–169 The interplay of woods and meadow can create dramatic effects.

IN THE FIELD

WEEDS AND THE ECOLOGICAL GARDEN

Old habits are hard to break. While I enjoy working with clients who are horticulturally knowledgeable, at times they are hesitant to carry out the horticulturally "wrong" techniques that I recommend, even when they require less work. I can recall one client in particular with whom this was the case.

Rose knew plants quite well and had created very sophisticated ornamental gardens at her previous home. She had decided to take a native approach at her new house, and I helped her plant a landscape that included a native garden at the front entrance. The property was small, but we had planted the vast majority of it, so there was a fair amount to tend. Rose managed most of it herself.

One day, we were standing in the front entry garden, which was then three years old. A lush expanse of woodland wildflowers including heartleaf foamflower (*Tiarella cordifolia*) and American alumroot (*Heuchera americana*) was close to filling in and creating an unbroken carpet. Excellent gardener that Rose was, there were almost no weeds. Now that the ground layer had become dense and competitive, I advised her to stop pulling weeds by the roots and start cutting them at ground level instead.

Weeding without thoroughly removing the roots was diametrically opposed to everything Rose had learned about gardening. Why would I advise this? The answer lies in a term we have frequently visited in this book: disturbance. All gardeners create disturbance. The question is how much do we understand its effects, and are we using that understanding to make our lives easier?

Pulling roots disturbs the soil, and disturbing the soil activates seed germination that typically results in weeds. So when you pull a weed by the roots, you are initiating and perpetuating a never-ending cycle of weeding. Alternatively, cutting the plant creates no soil disturbance and generates far fewer replacement weeds.

But the plant will just come back, right? Yes, it will in a garden where open mulched spaces between the plants allow it to receive light the minute it resprouts. If, however, you cut it below the foliage of a dense, intermingled ground layer, there will be little or no light available, and the resprouting weed is not likely to break through into the sunlight. In addition, it has to compete with the extensive root systems of the plants that have not been cut and are continuing to photosynthesize and become stronger. A particularly vigorous weed may break through once, or even twice, but it will not likely survive beyond that.

I once performed an experiment and found that, on average, it takes the same time to cut four weeds at the base as to pull one by the roots (depending on the weed, of course). I don't think my scientific method would stand up to a peer review, but I still believe that you end up ahead if you cut. And no weed-generating disturbance has occurred.

In an ecologically stable landscape, the stability is made possible by low levels of disturbance, which provide few opportunities for outside plants to penetrate

Rudbeckia seeded into Rose's entry garden from the surrounding planted meadows.

A compact inkberry holly provides contrast to the airy herbaceous entry planting.

A dense ground-layer planting creates a state of relative ecological stability where few weeds encroach.

and alter the system. By advising Rose to cut rather than pull weeds, I was trying to steer the garden toward ecological stability. But she just couldn't do it. She was far too good a horticulturist to leave that root in the ground. Though the ground layer was now dense, more weeds were still breaking through than I would have expected. Why? From an ecological standpoint, her thorough and fastidious weeding was keeping the garden in a state of ecological instability and more weeds than were necessary were the result.

Seven years later, Rose and I were again standing in the entry garden: the weeds were far fewer, but she had continued pulling them by the roots when they were able to emerge. Now, though, the heartleaf foamflower and American alumroot were engulfed in a delicate cloud of pink and red, the flowers of foxglove beardtongue (*Penstemon digitalis*) and red columbine (*Aquilegia canadensis*), neither of which I had planted in the

Foxglove beardtongue was planted above a ledge near the house entrance. Over time this plant seeded itself into gaps in adjacent gardens.

entry garden but had on an adjacent rock outcrop. What was going on here?

For years Rose had removed the weeds and prevented them from producing seed. All the while, the planted native species had been distributing seeds by the millions, resulting in a seed bank where they were now the dominant players. As a result, when Rose pulls a weed now, the disturbance is less likely to produce another weed and more likely to produce a native plant like beardtongue or columbine.

Ironically, neither plant still exists on the rock outcrop where they were originally planted. A longer lived and more competitive fragrant aster crowded them out of that spot. Is that a problem? No. Both beardtongue and columbine are short-term colonizers that live on through the seeds they distribute before they perish. That is their life strategy. For Rose, that fact meant she lost five plants and gained fifty.

Is disturbance, then, a good thing or bad thing? In the early days of Rose's landscape her routine disturbance produced weeds, but now it produces native flowers. An ecologist would say that in a healthy landscape dominated by native plants, disturbance is good. It allows new individuals to enter the system and prevents what might be termed "group senescence," an overall aging of the plants within the garden. To avoid too much instability, however, disturbance events are most beneficial when small, scattered, and infrequent, like Rose pulling the occasional weed in her entry garden.

Now that the seed bank has been altered, disturbance yields native plants like columbine and beardtongue instead of weeds.

SETTING THE ECOLOGICAL PROCESSES IN MOTION

The time has come to start making things happen, to turn plans into reality. But keep in mind that what is detailed here is only a beginning. Unlike a traditional garden that you plant as a finished product, with an ecological garden planting is just a start. The garden ecologist sets out plants or seeds into the landscape to get a living process in motion, one that the gardener will guide but also be guided by. This means that the design process does not end when planting begins. Decisions are made as the garden evolves, the vision adapted as the garden adapts. Some changes will be encouraged as they occur, others discouraged.

Suppose, for example, that elderberry (*Sambucus canadensis*) sprouts from a seed deposited in your garden by a bird. This is a vigorous shrub, which can grow rapidly to a height of 10 feet. Perhaps, as it matures, the elderberry blocks a view from the garden into a meadow, so you pull it. Maybe, though, the elderberry partners with a sumac to frame and accentuate the view, so you let it grow. Or maybe, rather than just reacting passively to the elderberry's appearance, you decide to encourage it. When its fruits ripen, you harvest the seeds and distribute them in disturbed or open areas of the landscape to enhance the elderberry's spread. In this scenario, design, planting, and management all occur concurrently, as the gardener interacts with and responds to a natural occurrence.

Establishing this cooperative venture between gardener and nature is the focus of this chapter, and the tools and techniques that will be described are those that apply to a wide variety of landscape types.

Before you start working, however, take care not to bite off more than you can chew or, in this case, maintain. Thus, even if you can afford to plant the whole landscape all at

A view across the meadow intentionally does not reveal a detention pond in the distance.

After directing the turf mower operator to leave an 18-inch strip unmowed at the base of the wall, I examined the area for volunteer native plants. When a particularly desirable plant appeared, like this elderberry, I marked its location with flags. I then directed the mower operator to resume cutting up to the wall, skipping over wherever I had flagged, and the volunteer elderberry was released.

once, you may not wish to do so. That's because the maintenance a new planting demands is typically far more than the same area will require once the plants become dense and weed suppressive. As a result, it often makes sense to install a landscape in phases so that the maintenance needs don't crest all at once. Even the extensive removal of weeds without a plan to manage and/or plant the affected area can result in maintenance over-load. If you create a vacuum, nature will fill it, and what rushes in will most likely be more invasive species.

The first really large-scale project I undertook was a 30-acre meadow with approx-imately 2 acres of garden. We planted the whole area in one autumn and the following spring. For the next two years, the maintenance was over the top: the client had to hire crews to weed out Canada thistle and other invaders in the garden, and he was not pleased. By the third year, however, the plantings were knitting together, there were fewer oppor-tunities for weeds to insert themselves, and the maintenance dropped off dramatically. In ensuing years, the garden became so self-sufficient that one person's labor two days each week was all that was needed to adequately care for the 2 acres of garden plantings. This was a happy ending, but the path by which it was reached would have been smoother if I had better informed the client ahead of time. He might well have decided to install the garden in phases over a period of years, so that earlier plantings would have settled in by the time that later ones were demanding that peak of initial care.

PREPARING THE LANDSCAPE

Before setting trowel, spade, or backhoe to the landscape, you should complete several preparatory tasks, including marking desirable and undesirable plant species in the landscape and controlling existing weeds.

MARK DESIRABLE AND UNDESIRABLE MATERIAL

Most likely there are plants on your site that you want to save, and others that you want to remove. You should clearly mark the difference with flags, ribbons, or marking paint before any work has commenced. The more clearly this is delineated in the field, the more efficiently the work will be performed and the less likely mistakes will be made.

While doing this, you have to make decisions about the lengths to which you will go to save individual plants. So, for instance, you may have a tangle of invasive plants with an occasional specimen of a common woodland plant at ground level. Saving those plants may greatly complicate the job of removing invasive species, making it impossible, for example, to use an herbicide. In such a situation, you may decide to sacrifice the handful of moderately desirable plants to speed and simplify the process, especially if the desirables are of species common elsewhere on the property. If the desirable plants are relatively unusual or ones that are hard to establish, a colony of trilliums, say, then the extra work involved in preserving them is usually justified.

In a former pasture being prepared for meadow installation, a small patch of existing butterflyweed has been flagged to preserve it from being eradicated along with the weeds.

CONTROL EXISTING WEEDS

A garden ecologist may welcome some volunteer plants but clearly not all. Species that appear spontaneously and that are invasive and unattractive qualify as weeds. Controlling these is essential to establishing a healthy garden ecosystem. To eliminate existing weeds on your site, you may use any or all of the following methods: removal by the roots, shallow grinding, herbicide application, smothering, starving weeds by frequent cutting, controlling the soil seed bank, and avoiding tilling and soil amendments

Removal by the roots

This technique is most effective for removing turf. Turf's shallow roots can be skimmed off near the surface, effectively removing the entire plant mass with minimal weed-producing disturbance. To treat a small area, a spade may be used, and for a larger area, you can rent a walk-behind machine called a sod skimmer. Deep-rooted plants require deeper digging, and the disturbance this creates will result in a stronger weed flush. Some of the other options that follow may be more effective for deep-rooted plants.

The pros of removal by the roots are that it removes entire plants, reduces the possibility of re-emergence, and minimizes soil disturbance if surface skimmed. Cons are that it is good for shallow-rooted sod but is not effective for controlling deep-rooted weeds. Physically difficult to accomplish, this process generates considerable amounts of material that must be disposed of.

Shallow grinding

This method is performed by machine and is most effective for areas with a heavy component of weedy shrubs and vines. The best piece of equipment for shallow grinding is a forestry mower. This relatively new contraption consists of a rapidly spinning cylinder mounted on the front of a tractor. The cylinder is raised above the about to-be-devoured plant, and as it is lowered, rugged teeth disbursed throughout the cylinder grind it to sawdust. Unlike a brush-hog mower, which cuts the plant relatively smoothly, the forestry mower leaves the remaining stubs shredded and far less likely to produce the buds needed to regrow. It can also go below the soil surface, thus grinding some of the root system and leaving a relatively smooth planting bed of soil mixed with sawdust.

For herbaceous weeds, a similar effect can be obtained by running over the field with other equipment like a power rake, which can grind a shallow layer of soil and roots. This must be repeated numerous times over the season to be effective, as one or two attempts are unlikely to satisfactorily eradicate the target weeds.

The pros of shallow grinding are that it does not cause deep disturbance that would bring up buried weed seeds, it is efficient over large areas, and it avoids the use of herbicides. The cons are that it requires special equipment and is difficult to implement in tight spaces.

Herbicides

I do not take the use of chemicals lightly, but often a selective and temporary application of herbicides is the most effective method of control. In almost all cases, once the target vegetation is established and dominant, chemical use can be vastly reduced or eliminated entirely. Compared with some actions, particularly repeated deep tilling, herbicide use may even be the one with the least environmental impact. In large areas dominated by invasive plants, conversion to native plant communities may not even be possible without the use of herbicides.

That said, it is important for those who want to be more cautious than government regulations dictate—and I count myself in that group—to closely track independent research and stay current on which materials are considered environmentally benign and which have unacceptable impacts. This process is performed at our offices several times each year and has caused us to discontinue use of certain herbicides even though they are considered safe by the U.S. Environmental Protection Agency. In almost all cases, once the target vegetation is established and dominant, chemical use can be vastly reduced or eliminated entirely if desired.

You may infer from my comments that I resist tackling large weedy areas when herbicide use is off of the table. Actually I welcome the opportunity. I don't use chemicals because I love them; I use them because often they are the most effective method. While approaching these types of sites without herbicides is certainly more complicated and usually more expensive, we will never learn how to do it better and cheaper if we don't try. As long as my client understands the ramifications of an alternative approach, I welcome the opportunity to further develop effective approaches and have done just that on many occasions.

Methods of applying herbicide include:

- broadcasting the material over the vegetation, a technique reserved for situations where most or all of the vegetation is undesirable,

- spot spraying undesirable plants, when just a few bad actors are intermingled with much desirable vegetation, or,

- with woody weeds, cutting off the stem and painting the cut end immediately with herbicide, a technique that is most effective when practiced in late summer, autumn, or early winter.

Flags in this woodland mark invasive species to remove. To avoid disturbance to the soil and the likelihood of resulting weeds, the flagged plants will be cut at the base and the cut stem painted with herbicide. This targeted application has no impact on nontarget vegetation or the environment.

Types of herbicides include nonselective products that kill any vegetation with which they come in contact and selective ones that kill only certain classes of plants. The latter includes broadleaf weed killers that target broadleaf plants but do not harm grasses and grass-specific herbicides that will not harm broadleaf plants. Selective materials can be particularly useful where desirable and undesirable plants are both present. A nonselective material would kill both, and throw the baby out with the bathwater. If the desired plant

top This 1-acre field has been covered with black plastic to smother the weeds. It will remain in place for one growing season prior to seeding.

bottom The same field now features a seeded meadow in its second year of growth, with short-lived species like blackeyed Susan and tickseed visually codominant. The plastic was later moved to the field beyond in preparation for more seeding.

is a native grass like purpletop tridens (*Tridens flavus*) and the weed is a broadleaf species like common buckthorn, using a material that targets only the buckthorn will save the baby and the three years it takes to grow purpletop tridens from seed. In a woodland where the highly invasive Japanese stiltgrass (*Microstegium vimineum*) is mixed with broadleaf wood asters and spotted geraniums, a grass-specific material will target and control the weed but preserve the valued wildflowers.

Organic herbicides use vinegar and other, sometimes quite caustic, natural materials to control weedy plants. These organic herbicides are not systemic; that is, when sprayed on foliage they kill the leaves but do not travel through the plant to kill the roots as well. Consequently, they generally require multiple applications at short intervals to be effective. Most organic products are not selective and will injure desirable plants as readily as weeds.

The pros of herbicides are that they do not cause disturbance to the soil and so do not promote a subsequent flush of weeds. Herbicides are often the most effective methods available from both eradication and cost standpoints, although organic herbicides are generally less effective than their synthetic counterparts. The downside to herbicides is that, even with research tracking and discriminating materials selection, some herbicides may have environmental impacts that have yet to be determined.

Smothering

Once applied, smothering materials such as black plastic, newspaper, and cardboard must be left in place for a full growing season to a year to be effective. This technique is generally reserved for small areas, and a 1-acre field is the largest I have planted this way. On that project, after smothering the existing vegetation, the plastic was removed, the meadow seed mix was planted, and the plastic was relocated to smother an adjacent acre.

The pros of smothering include the lack of soil disturbance, which provokes the germination of dormant weed seeds, and no use of chemicals. This technique is very effective for preparing an area for planting. The cons are that this method of weed killing is expensive and time consuming. In addition, when using black plastic, it involves the use of a nonrenewable resource.

Starving weeds by frequent cutting

This technique is most effective with annual and biennial weeds. It involves cutting the weeds and undesirable plants back to the roots frequently and consistently so that they never have an opportunity to photosynthesize and replace the resources drained from them by the cutting. Consistency is crucial: if you allow the plant to regrow and retain leaves for a significant period between cuttings, this technique will not work. How long this will take, of course, depends on the weed.

The pros of sequential cutting are that it produces no soil disturbance and does not require the use of chemicals. The cons are that this technique may require a prolonged period of cutting when applied to perennials. In addition, it may not work on weeds with large capacities to store food in roots or rhizomatous or tap-rooted species, such as mugwort (*Artemisia vulgaris*) and Japanese knotweed. Cutting is not a good method for trees that spread clonally by rhizomes. These plants, including Oriental bittersweet and tree of heaven, expand more vigorously when cut.

Controlling the soil seed bank

Even if soil disturbance has been avoided, eliminating the existing weed cover will remove competition that had previously suppressed weed seeds that are near the surface. To eliminate these as well, you may allow the seed bank—the dormant weed seeds on the site—to emerge and then eradicate those emerging weeds before planting. Ideally, you should allow both a warm and a cool season (a summer and a spring or autumn) to pass so that both cool-season and warm-season weeds have an opportunity to germinate. Killing such newly sprouted weeds is easily accomplished with an additional application of some suitable herbicide. If, however, you wish to avoid the use of chemicals, you can instead lightly rake the area, either by hand or with a York rake, and uproot the seedlings when they are newly emerged and still easily uprooted.

You should limit the importation of any additional weeds to the site by avoiding bringing in any soil, compost, or other seed-bearing materials. In addition, it's important to specify that contractors clean the tires of their equipment before they bring it onto your site. Otherwise, they may bring in weed seeds in soil caught between the treads.

Carefully inspect any plants brought onto the site for weeds in the root ball or container. This may sound like excessive caution, but I have seen otherwise. On one project I undertook, I found an infestation of mugwort, a particularly aggressive and stubborn weed, in one area. Wondering how this weed had found its way out into fairly remote woodland, I inspected the area and found that the epicenter of the infestation was a silverbell (*Halesia* species) tree planted before I undertook the job. In its root ball I found mugwort.

Avoid tilling and soil amendments

These operations are so common in conventional horticulture that many gardeners think that they are an inevitable part of any landscaping project. My advice is to avoid both as much as possible.

By turning the soil, tilling brings to the surface weed seeds that were formerly too deeply buried to germinate. In this way, tilling encourages the immediate growth of a whole new crop of weeds. It also opens the site to erosion by removing the matrix of roots that helped stabilize the soil surface. In addition, tilling upsets the delicately balanced soil ecosystem, destroying beneficial soil organisms.

Fertilization typically benefits opportunistic weeds more than thrifty natives. Indeed, if you have chosen your plants well to match the characteristics of the site, you should not need to fertilize.

When planting a meadow, do not incorporate extra organic material such as compost into the soil. The meadow species won't benefit from higher levels of organic material, and with it are likely to come weed seeds. Woodlands can benefit in certain cases, which will be discussed in the chapter "Techniques and Tools: Woodlands."

PLANTING THE LANDSCAPE

This is when all the preparation and planning come to fruition. As in other aspects of ecological gardening, there are some special considerations.

DECIDING WHEN TO PLANT

As in traditional horticulture, the best planting season in most of North America is spring and autumn. The exception is seeded meadows, where the planting season can extend into early summer due to the adaptation to hot weather of native prairie grasses and flowers.

Some timing considerations are unique to native design. Where establishment watering may be difficult because of the remoteness or scale of the planted area, it can be minimized by planting in early autumn. That's because plants set out then have two cool growing seasons—autumn and spring—to infiltrate their roots into the soil before they experience the heat of their first summer.

If you plant plugs, particularly shallow-rooted ones, it's also best to plant in early autumn. Not only does this reduce watering requirements, but, by allowing the plants to root in before the onset of winter, it prevents frost heave from pushing the plants out of the ground.

PURCHASING WOODY TREES AND SHRUBS

Nursery professionals strive to produce trees and shrubs of uniform size and with thick foliage all the way from the top to the ground. That is not the way most plants grow in nature. Rather, this artificially lush look is a sign that the plant has been babied, spoon-fed with fertilizer and water. Put such a plant out in a natural setting, and it's going to suffer a bad shock, so that it struggles to survive or even dies. This is especially true in a naturalistic setting where irrigation is minimized during the period immediately following installation. For this reason, it is better to select more sparsely leaved, modest-sized, thrifty plants, which can focus their energy on spreading their roots into the soil rather than on a battle to support an unnatural quantity of foliage.

Whoever said that seeing the stems of a shrub is an affront, anyway? I wonder if this is a legacy of Victorian

An overly uniform and bushy sweet bay magnolia would look out of place in this wild garden.

MIMICKING THE LOOK OF NATURE

The uniform look that most nursery professionals regard as desirable is actually undesirable in a naturalistic setting, particularly in the case of trees and shrubs. In a natural woodland or shrubland, woody plants vary in size and shape. When you interplant with perfect nursery-grown stock, the result is visually discordant, as the new plants obviously do not belong.

A perfect illustration of this occurred years ago when I was working on a project with a new contractor. He looked at the airy shrubs and irregularly branched trees delivered by the nursery—plants I had selected—and pronounced them "junk."

"I can't believe," he added, "you're selling these to the client."

The plants were healthy. What bothered him was their non-uniform, thrifty character. Later, though, when the planting was done, he changed his opinion.

"Yeah, I get it," he said. "It looks like the woods."

Not only did these plants have a more natural appearance, they were hardier. Though they were scattered throughout an extensive area of woodland and received minimal nurturing after planting, they mostly flourished.

That contractor and I installed many subsequent plantings together, and his company went on to adopt a strong native plant focus.

gardeners with their horror of exposed legs? In fact, so-called leggy woody plants not only have a more natural look, they leave more room for ground-layer herbaceous plants. Native azaleas, for example, are frequently criticized by gardeners as being too open branched, but that's in part because gardeners do not plant them as nature does, with ferns and wildflowers coming up through them; the combination sets off both the shrubs and the herbaceous plants. Far from an aesthetic liability, the stems of a shrub like highbush blueberry (*Vaccinium corymbosum*) can display a twisting, bonsai-like architecture when mature and even provide a focal point in the landscape.

Most gardeners like to plant large specimens, yet the truth is that smaller plants can affect more area with less expenditure, which can be important if you have decided to convert a large area of lawn to shrubland. Consequently, replacing large balled-and-burlaped shrubs with 1-gallon containerized plants can turn an out-of-reach pipe dream into an affordable project.

It only takes a tree or two to create some shade in a back yard, but creating a woodland takes a lot of trees. Large trees and the machines needed to move them can run up large landscaping bills very quickly. In addition, research published by W. Todd Watson showed that small trees suffer significantly less transplant shock and often catch up. That said, when planted in wilder areas, very small sapling trees can be difficult to locate and maintain, unless flagged. Midsize potted trees of approximately 6 to 8 feet can serve as a good compromise.

Well-grown, deep-rooted landscape plugs have extensive root systems that allow them to establish readily. Given their small size, they can be purchased at far less cost than plants in quart- or gallon-sized containers.

PURCHASING HERBACEOUS PERENNIALS

Whether in the meadow or the woods, herbaceous plants form the most important element of the ground layer, and the ground layer serves as the most important line of defense against weed incursion. While seed is always an option for establishing ground-layer compositions over large areas, the desire for quick establishment or a more gardenesque appearance may require the use of live plants, again in relatively large quantities.

Native perennials are often grown as plugs in trays of individual cells, similar to the way exotic evergreen groundcovers like pachysandra or vinca are packaged. Plugs can be a good way to cover more real estate for less money. Although the original cell packs were shallow, so that the plants in them were shallow-rooted and difficult to keep alive after planting, nurseries have recently taken to growing plugs in deeper cells, and the more extensive root systems have vastly improved the plants' survivability. Use the latter whenever they are available.

opposite Bare lower branches allow a shrub's architectural form to remain visible, accented by low-growing herbaceous underplantings.

For a small area, larger potted plants may be more economical than plugs even though they are individually more expensive. This is because you must purchase plugs in trays of thirty-two or fifty plants, and that may be more than your area will accommodate

above Newly establishing plants in a mulched area are particularly susceptible to deer browse, as they stick out like lollipops. To avoid this situation, phlox has been planted in the same hole as Pennsylvania sedge, a species that deer typically avoid.

right As this deer-favored white wood aster expands into the landscape, its protector, a colonizing fern species, expands with it. It is important that the two plants are of similar competitiveness so that one does not outcompete the other.

for any one species. This is especially true when planting large, robust plants such as cutleaf coneflower (*Rudbeckia laciniata*), which may grow to a height and spread of 6 feet or more. Just a few plants like this may fill all the space available.

SOIL AMENDMENTS

Traditionally, gardeners have recommended incorporating many different amendments into the backfill soil around the root ball of new plantings. These usually include organic matter and fertilizer in various forms. I avoid these, because research has shown little long-term benefit from such additives.

Also making their way into the marketplace are biological additives, including compost tea, micronutrients, biochar, and vitamins. In some cases their benefits are debated and controversial, and I have not used them to any significant degree. Mycorrhizal fungi, however, are essential for some specialist species and are often included in the seed packet for those species. This additive does appear to be beneficial. In some cases, mycorrhizal fungi can also be mixed with soil in the planting hole for species with known fungal relationships. Oak trees are known to have particularly common fungal associates, some of which may no longer be present in the soil, particularly if oaks have not grown on the site for many years. Whereas additives like fertilizer or compost remain in the hole and may discourage root expansion, the fungi attach to the roots and move into the surrounding soil with the host plant, providing permanent benefit.

DEER PROTECTION

I have noticed over the years that plants in a meadow rarely suffer from significant browsing by white-tailed deer. Even the plants that deer favor seem to escape this form of attention when intermingled with plants that the deer don't eat. Apparently, the deer don't thrust their noses down through something inedible to find a tidbit. I have applied this

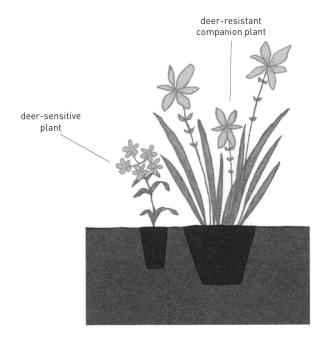

deer-resistant
companion plant

deer-sensitive
plant

Planting a smaller specimen
of species favored by deer in
the same hole as one disfa-
vored can afford protection
to the susceptible plant.

observation to the planting of herbaceous perennials. When planting something I know the deer relish such as white wood aster, I also plant in the same hole something that they will not eat, such as common ladyfern (*Athyrium filix-femina*). Just as in the intermingled meadow composition, the deer can't get at the plant they like without also encountering the one that they don't. It is important when doing this to match plants of a similar level of aggressiveness so that one doesn't overwhelm the other.

Deer are another argument in favor of planting midsize instead of small seedling trees. I like to plant specimens sufficiently large that I can prune up the branches to a minimum of 4 feet, which is higher than deer ordinarily browse. Another danger from deer is that bucks rub their antlers and foreheads against the base of young trees during mating season to signal their presence to other deer, and this will damage the bark and affected layers beneath it. To prevent this, I insert three pieces of rebar (concrete reinforcing bar) into the ground equidistant around the trunk of a young tree. This is easy to install in early autumn prior to mating season, confers protection against the rubbing, and is inconspicuous, especially once the rebar is covered with rust. The rebar is also easy to remove after rutting season has ended. It can be stored efficiently over winter and reused the following autumn if the tree is still of a susceptible size.

Unfortunately, I don't know of any silver bullet for preventing deer from browsing shrubs. In their case, the only effective protections are applications of repellents or deer exclosures. If you use repellents, try switching products from time to time. In my experience, the deer eventually become accustomed to any one repellent and will then ignore it.

Only one clump of invasive purple loosestrife (*Lythrum salicaria*) exists in this naturally occurring meadow of all native species. Removal of the loosestrife should be prioritized over areas already largely dominated by invasive species.

MANAGING THE LANDSCAPE

An important concept when managing any landscape is to work from best to worst. That is, if you are confronted with two areas, one that includes just a sprinkling of weeds and another that is overrun with invaders, it makes sense to weed the area with fewer weeds first. This may seem counterintuitive, but if you were to tackle the badly infested area first, you would find yourself investing lots of time in that, and by the time you return to the relatively better area, the situation there is likely to have deteriorated so that what was a small job is now a big one.

CONSIDERING INITIAL FLORISTIC COMPOSITION

The theory of initial floristic composition proposed by the ecologist Frank Egler states that much, if not most, of the vegetation present on a site at any given time was also present or entered the site when it was last opened up by disturbance. When a disturbance event occurs, whether a tree blowdown, the final cultivation of a farm field before abandonment, or the scraping action of a bulldozer, the site is open for business and

welcoming all comers. Once vegetation has been re-established, planted or otherwise, it becomes very difficult for newly introduced species to penetrate. When a bird or the wind deposits a seed, the chance that it will result in a surviving plant or germinate at all is very low.

This means that you should count on devoting the most attention to your landscape while it is in its early stages after disturbance from clearing or weed eradication. By selecting for planted species and against weeds during the inevitable sorting out period, you are directing the vegetative trajectory toward the favored species. Once these become dominant, weeds will be few and far between and management requirements will drop dramatically.

Nearby, invasive mugwort dominates a field, and removing it will take a significant amount of time. By attending to the mugwort first, removal of the purple loosestrife will morph from a little job into a big one.

EXPLOITING THE DIFFERENCES

An overriding principle in managing for desirable and against undesirable species when they are both present in the same area is to exploit the species' differences. If you want to promote or discourage any particular plant or group of plants, identify how those plants differ from those around them and then make use of that difference.

Three rebar stakes have been placed around newly planted trees in this woodland to protect the plants from buck rub. As the rebar rusts, it blends into the surroundings better than wooden stakes. Once the trees reach an adequate size, the rebar can be removed.

For example, when you mow high to control tall weeds, set your blade higher than the tops of desirable plants to exploit a difference in heights. Mow or spray in early spring to control cool-season weeds without harming still-dormant warm-season plants to exploit a difference in the plants' growth schedules. Many of the management techniques found in the remainder of this chapter and those that follow will make use of this strategic approach.

WEED CONTROL

When managing the weeds in your landscape, you need to know the life cycles and reproductive strategies of the weeds on and nearby your site and then select the most appropriate control technique for those species. Various weed control measures are outlined here, including some that many gardeners may be unfamiliar with.

Pulling weeds

Gardeners have been pulling weeds for as long as they have been gardening. But is it really the most efficient approach? As I describe in Rose's story at the beginning of this chapter, pulling up roots can be a disturbingly counterproductive weed stimulator. Once a dense ground layer has been established—and that should always be one of your primary goals—it's better to reach through the desirable plants and cut the weed at ground level. It will likely resprout, but now you have placed it at a competitive disadvantage. The

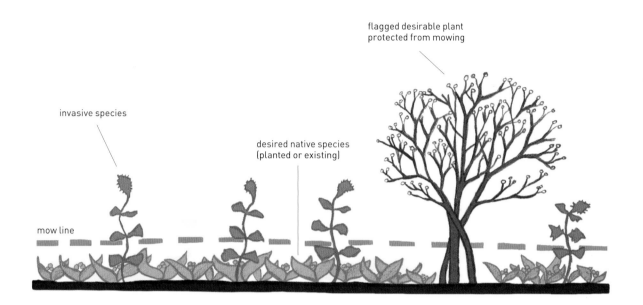

flagged desirable plant
protected from mowing

invasive species

desired native species
(planted or existing)

mow line

In selective height cutting, taller weeds are cut over the top of low-growing desirable species, allowing them to continue photosynthesizing while the weeds are forced to expend their energy regenerating.

resprout has to start at ground level, underneath the uncut vegetation where light availability is vastly reduced. What's more, the weed is now operating at reduced efficiency, having wasted resources replacing lost foliage. Meanwhile the planted species are motoring along in their preferred habitat, photosynthesizing without interruption. The weed may break through once, or even twice, but it takes much less time to cut a weed than pull it by the roots. You're likely to still end up ahead and without disturbing the soil and waking the weed seeds.

Of course, I never say never. When you are up against the most pernicious invasive species, the woody vine Oriental bittersweet, for example, and when spot application of herbicide is off the table, good old root yanking may be required because cutting simply promotes rampant resprouting. But for the most part, cut the weed and let the remaining desired plants finish the job.

Selective height cutting

This practice is another example of exploiting the differences between the desirable and undesirable plants. If most of the weeds are taller than the desirable species, you can disfavor them by mowing with a brush-hog or string-trimming them just over the tops of the desirable plants a couple of times annually. This tends to be an especially effective strategy in woodland settings, where the ground-layer of ferns and wildflowers is often low. Of

top Periodically weed whacking the emerging blackberry over top of the Canada mayflower will favor this short native wildflower.

right Over time, this procedure resulted in a solid carpet of the mayflower. When the cutting crew encountered the taller ferns, they raised their equipment to cut over the top of them as well.

bottom When an oak tree seedling emerged, it was flagged so that was not inadvertently cut along with the tall weedy growth.

course, the weed won't expire the first time you do it, but having to continually regenerate while its competitors have not will place it at a competitive disadvantage that causes it to decline over time.

You can exploit this distinction even more deliberately by designing to create a difference in plant heights. That is, you can select low-growing species to plant among taller weeds and then apply the selective mowing to foster the plantings at the weeds' expense. This approach can allow you to begin introducing native species without first denuding the site. You must consider the relative aggressiveness of the native species you're planting and the weeds you're planting them among; your plantings must be able to hold their own. Selective height cutting must be continued until the desired plants are adequately outcompeting the undesired species.

Seasonal cutting

This practice involves mowing the weeds when they are growing actively and when the desirable plants are dormant. Thus, if you are battling cool-season weeds invading a meadow of warm-season grasses and wildflowers, then mowing one or more times as necessary in the spring, when the cool-season weeds are active and the desired species are not, provides a convenient means of weakening only the undesirable species. I used this strategy to convert a thicket of shrubs to a field of native little bluestem in Dutchess County, New York. It is particularly useful to favor native warm-season grasses over the

European cool-season grasses that dominate most open fields in North America. Thus, timed cutting is yet another example where exploiting a difference, in this case growth period, can have a large effect with very little effort.

Do nothing

Not every weed demands action. If the invader is a short-lived species that follows on the heels of disturbance, such as Pennsylvania smartweed (Polygonum *pensylvanicum*), and the desirable plants are long-lived perennials characteristic of a later stage of succession, such as pink turtlehead (*Chelone lyonii*), then you can simply let nature take its course. The natural process of succession will eliminate such weeds without your intervention.

Using herbicides

It is generally easier to avoid herbicide use when managing large-scale planted landscapes than when preparing the site for them. That said, herbicides remain a very useful tool, and given that their use in management is generally restricted to targeted spot applications to individual weeds (not broadcast over entire planting areas), the quantities of material used are very small.

Spot application of herbicides with a backpack sprayer is a way to control troublesome invasive plant species while minimizing the amount of chemicals applied. A wick applicator, which wipes the liquid herbicide right onto the targeted plant, is even thriftier and prevents any of the material from reaching the ground. This tool can be very useful where weeds are intermingled with desirable plants, as it allows for a very precise application.

You can also use cutting to create a height differential for spot herbicide application where desirable and undesirable plants are intermingled. Cut the weeds down near the base and when they re-emerge, they will be lower than the desirable plants. If you then spray the herbicide near the ground, you will hit the foliage of the undesirable plants and not that of the now taller, desirable ones. In this case you have created a difference to exploit.

Timing the application judiciously plays a crucial role both in effective control of the targeted weed and in avoiding injury to other vegetation. Japanese knotweed, for example, another particularly pernicious invasive, will quickly resprout from the roots after an herbicide application unless it is applied in late summer and autumn. Timing can also protect desirable vegetation. For example, an application of the appropriate herbicide in spring in a field of little bluestem can control the targeted weed with little harm to the bluestem, which is a warm-season grass and so dormant and less vulnerable during a spring application.

Broadcast applications of herbicides are not generally a part of landscape management—they are more often used to prepare the landscape for planting—but they can be useful in special circumstances. If, for example, Japanese stiltgrass, a particularly aggressive exotic annual grass, invades a woodland with an understory of broadleaf wildflowers, you might broadcast an herbicide designed to kill grasses only, leaving the wildflowers unaffected. In designing for such management, grasses would have been excluded to allow this technique to be used without damaging any of the planting.

Seedlings of blackeyed Susan and foxglove beardtongue are emerging amid a sea of invasive Japanese stiltgrass. The stiltgrass was documented during site analysis, and so only sedges and broadleaf flowers were planted, allowing grass-specific herbicides to be applied. Once the desired species become dominant, the herbicide can be eliminated and native grasses can be added to the composition.

While on the subject of designing for management, I should mention the advantages of planting a groundcover of Pennsylvania sedge, or any sedge for that matter. Because sedges are neither grass nor broadleaf, you can use a broadcast application of herbicide to selectively control both, which are virtually all the weeds there are. As with herbicide in general, discontinue use once the favored plants are dense and no longer need the help. At that point you can even add in species other than sedge, if desired.

WATERING

The goal for irrigation should be self-sufficiency. That is, you want to keep new plantings alive until their roots establish, but you don't want to foster such lush growth that the plants become addicted to supplemental water.

Achieving this requires several things. To begin with, you should time your planting properly so that it occurs at the beginning of a period of relatively moist weather. With the exception of heat-adapted meadow plants grown from seed, it's best not to plant at or right before the onset of summer with its heat and droughts. Hydrogel formulations (like Agri-gel) applied to the roots to absorb and hold moisture are somewhat controversial. Some research suggests that they hold the water so tightly that it isn't available to the surrounding plants anyway.

A temporary irrigation system of sprinklers or soaker hoses can be useful for helping new plantings through periods of insufficient rainfall until the plants can extend their roots down into the soil. Apply water sparingly though. One thorough irrigation that moistens the soil deeply is better than repeated wettings of the surface soil. Deep watering encourages deep root growth, whereas shallow watering fosters weeds more than the desirable plants, and plantings can succumb to that competition as surely as they will to drought. Above all, don't apply more water than the plantings require to survive, because the extra will only go to nurture weeds.

A permanent irrigation system such as in-ground sprinklers will not be necessary if you have chosen plants whose needs match the moisture regime on the site. Your choices, then, should be informed by the regional climate, any microclimates on the site, soil type, topography, and the depth of the water table, all factors considered during site analysis and plant list development.

Even in periods of drought, you can do more harm than good with ill-considered irrigation. Upland meadow plants have the ability to go semi-dormant during moderate droughts; they don't wither, but they do stop growing and their water needs drop. Irrigating plants in this condition will bring them back to active growth, and make them more vulnerable to drying out. So if you water your meadow during a dry spell, be prepared to continue watering until the drought ends.

SOIL AMENDMENTS AND MULCHING

I generally advocate choosing plants that favor existing soil conditions in place of altering soil conditions to favor the plants. There are exceptions, however, including meadow plantings that are to be planted in highly fertile soils. Of course, this may be the situation where a traditional gardener would be *least* likely to amend. I'm not trying to be a contrarian here, but rich soils are generally rich in weeds, and many upland meadow species tolerate infertility quite well. Applying sulfur to the soil lowers the soil pH, which makes nutrients in the soil less available to plants, essentially mimicking infertility and favoring the meadow species. Additional exceptions will be discussed in the chapters that relate to the specific habitat types to which they apply (meadows, shrublands, and woodlands).

Bark mulch is one of the most overused materials in contemporary gardening. Although it can be a useful temporary cover as the ground layer becomes established and fills in, it is generally impractical and counterproductive when used to blanket larger areas. It's better to employ a sort of living mulch by sowing short-lived, fast-growing plants between the more long-lived plantings. Eventually the long-lived species grow together and the seeded species fade out of the picture, much as they would in the wild, where early-stage species drop out as the process of natural succession marches on.

Don't ever use landscape fabric. As the mulch with which you hide the fabric begins to decompose, it becomes a nursery bed for weeds whose roots become entangled in the fabric below. If you then pull out the weeds, they bring up unsightly little humps of the black fabric throughout the planting.

Raw wood chips are held in disfavor as mulch by traditional horticulturists because they temporarily bind up nitrogen as they decompose. This doesn't disturb me, as ecological landscapes don't depend on the sort of excessive fertility favored in conventional gardens. In fact, a study by Rutgers University found that mulching a site with raw wood chips promoted a spontaneous recruitment of volunteer native trees, principally the sort of late-stage succession trees such as oaks and hickories that form the canopy of a mature woodland, and not the pioneer trees like birch and poplar that normally precede them. This

top Instead of using mulch to suppress weeds, short-term species were seeded into this garden amid long-lived species installed as container-grown plants.

bottom As the late-stage species matured, they outcompeted the short-term seeded vegetation. In the event of disturbances, seeds deposited by the short-term species could germinate and fill in the gaps until long-lived growth once again takes over.

effect seems to be due to the fact that the wood chips mimic the litter of branches, bark, and fallen trees that normally cover a forest floor. In this case you are skipping over the pioneer woodland stage that would normally precede the advent of late-stage species by mimicking the ground layer conditions in which they normally would germinate. When we converted shrubland to grassland on the Dutchess County, New York, property, we moved succession backward. The application of wood chips allows us to fast-forward succession to achieve the target plant community more quickly, in this case an oak-hickory forest.

Small gravel, such as ¾-inch round pea stone, will promote the germination of dry ledge and scree species such as columbines (*Aquilegia* species) and many alumroots (*Heuchera* species). As with raw wood chips, you are mimicking the conditions in nature that stimulate germination of their seeds. Plants in this category often seed prolifically under the right conditions; planting a few can result in hundreds more, but only if you provide that condition.

MANAGING PLANT PROLIFERATION

This can take two forms: encouraging proliferation if the plants in question are desirable or discouraging proliferation if the plants are unwanted.

Assisting seed dispersal

Seed dispersal can be assisted by collecting seed of desirable plants when it is ripe and then distributing it into gaps, either natural and existing or created, in the vegetation on your site. Different types of plants ripen their seeds according to different schedules. For information on when a particular species ripens its seed, you can consult seed catalogs and online sources. It's also good to learn to recognize ripe seeds in the field, as seasonal climate variations and geographical location can cause ripening time to vary from year to year or place to place. Many botany texts can prepare you to do this. Be sure to scratch up or otherwise disturb the soil where you disperse the seeds, whether by sprinkling them over the soil or just depositing stalks with seed heads.

Sample seed dispersal calendar of species that disperse by gravity

Botanical name	Common name	April	Early May	Mid May	Early June	Late June	July	Aug	Sept	Early Oct	Late Oct
Aesculus parviflora	bottlebrush buckeye						flowering		fruit + dispersal		
Aquilegia canadensis	red columbine	flowering + fruiting						dispersal			
Heuchera americana	alumroot		flowering	fruiting	dispersal						
Lobelia cardinalis	cardinalflower							flowering		fruiting	dispersal
Penstemon digitalis	foxglove beardtongue			flowering		fruiting		dispersal			

Formulating a calendar of when seeds are ripe for dispersal can guide the timing of your proliferation assistance efforts.

Preventing seed dispersal

Of course, there are plants in the landscape whose proliferation you would prefer to prevent. Some of them you will remove by the methods so far discussed, but if you can't get to that right away, at minimum prevent seed formation by cutting them at the appropriate time. In the case of many annual and biennial plants whose life span is limited and dependent on seed dispersal, this may be all that's needed to control their spread. Don't pull these plants. Their life cycle is short by definition, and the plant itself will expire without your help. Pulling just creates disturbance and sets the stage for a never-ending cycle of more weeding.

Assisting plant proliferation by vegetative means

Some plants reproduce mainly vegetatively, spreading by rhizomes or stolons, and don't set much seed. You'll be more successful in spreading species of this sort by removing a living piece of a plant—a rooted stolon (a creeping stem) or rhizome (a creeping root) with stem or stems attached—and transferring it to the soil in another location. This can be a valuable technique for filling gaps in your planting and forestalling weed invasion. Be sure when you harvest the plant piece that you take it from a spot where the plant in question is intermingled with other ground-layer plants. There's no point in creating a weed-prone gap in order to plug one in another location.

opposite and above Sometimes spreading seed is exceedingly simple. To establish cardinalflower, place a seed stalk on the surface of the soil. Don't bother scraping it in, as the seed of this species requires light to germinate. With this simple method requiring only a few minutes of effort, I had a new population emerge and flower two years later.

DESIGN DURING MANAGEMENT

Just because you have reached the management phase with your landscape doesn't mean the design process has ended. A healthy, ecologically managed landscape will readily sprout new volunteer plants. Your job as the designer is to decide whether to capitalize on this by leaving any given plant to grow or to remove it. Frank Egler called this process intaglio, a form of printmaking in which the image is created by cutting into the surface of a printer's plate and removing material to leave an incised image. Another way to think about this is as landscape editing. This is an exciting aspect of the ecological garden: the landscape keeps developing, continually presenting the gardener with new opportunities.

CRITICAL MASS

Achieving critical mass is the payoff that comes after a number of years of conscientious management. You've succeeded in removing the weeds that spread by vegetative means, and you've kept the other weeds from setting seed. Meanwhile, you've successfully fostered desirable plants that have spent years distributing seed throughout the landscape.

Mowing a path around a colony of gray dogwood allowed the patch to be preserved but prevented its suckering growth from further invading the larger meadow. Leaving the tiny meadow clump unmowed was purely an aesthetic touch.

As a result, the seed bank has become dominated by the native species you have planted or managed for. Now when a disturbance occurs—when a shrub or tree dies, when the dog runs out and digs a hole in the garden—what emerges is most likely to be a desired native rather than a weed.

I have been bad-mouthing disturbance through much of this book, but it isn't always bad. Originally, it was a threat, but now that the seed bank has been altered, disturbance is something you can welcome. Ecologists define the ideal disturbance pattern as small events, scattered throughout the landscape. These limited, continual disturbances are just part of the landscape's evolution. It's the way in which the ecosystem replaces aging, senescent plants, the way in which the landscape renews itself, while presenting you with design opportunities. Now you've restored the landscape to ecological health, and when the dog scratches around in the garden, good things are likely to happen.

top Species from a nearby meadow have seeded themselves into a garden closer to the house, blending the two such that there no longer seems to be a division between cultivated and more natural areas.

bottom Sometimes plantings on your property have impacts beyond the property lines. The purple-flowering ironweed growing in this field came from plants in a meadow I designed for a neighboring property.

Meadows play many important ecological roles from stormwater retention to wildlife habitat, but their sheer beauty can also take your breath away. Rose's meadow proved no different.

top A view across the meadow intentionally does not reveal a retention pond in the distance.

bottom A look in the opposite direction underlines a view of an off-property barn.

top Rose's house recedes in the distance as we head down the meadow path bordered by Indian plantain, purple coneflower, and a mix of yellow composite flowers.

bottom The meadow on the right belongs to Rose. After it was well established, we planted a meadow for the neighbor on the left.

Next we planted a meadow at the far property beyond the break in the trees, completing the scene.

A DO-NOTHING ATTITUDE

The kickoff meeting was well attended. The property manager, landscape contractor, general contractor, landscape architect, and the owners of this 450-acre estate were all present and accounted for. The house was still under construction, so the meeting took place in a cramped construction trailer around a rickety make-shift table. While the quarters were modest, the 40-acre meadow that I was about to plant was not. None of the players at the meeting, save the owners who had done quite a bit of homework, had a clue how the seeding would occur, and they had plenty of questions.

"Should we first bring in topsoil?" Although the existing soil was bony and infertile, my answer was "No."

"Should we fertilize?" "No."

"Some compost at least?" "No."

"Do you want the field to be tilled?" "No."

"What kind of irrigation do you need?" "We don't need any irrigation."

The seat was getting hotter, and then came the question that probably confirmed my incompetence in their eyes. "When do you want to plant this thing?" "July."

They looked at me across the tiny table. I could see their expressions—actually I could *feel* their expressions—and I knew pretty much what they were thinking. "Sure, you're the golden boy now, but if this becomes a weed field, you'll be out of here faster than a dandelion puffball in a hurricane."

They were right to be skeptical. They had probably seen more than a few well-meaning, quick-fix meadows flame out after a year. I was taking a more thoughtful approach, but they had no way of knowing that. So I explained my rationales.

Regarding the addition of topsoil, I had selected species that were derived from habitats with similarly infertile soils. This is where they normally live, and they are perfectly happy with the soil as it is. Nice rich topsoil would have only vastly improved prospects for invasive weeds. In addition, importing new soil often imports new weeds, and we had enough of those on the property already.

I made a similar argument for avoiding fertilizer, compost, and irrigation. If the meadow plants were living in a vacuum, like a pot in a nursery or an isolated spot in a garden bed, they would likely benefit from some extra water and nutrients. But in the meadow they are contending with the element of competition, where survival is achieved by whichever species can operate in the environment most efficiently given available resources. Adding resources may appear to be helping all of the players, but it mainly favors those that most need those resources. And in this environment, that would be the weeds.

And why not till? When the folks around the table see a freshly tilled field, they see a nice smooth planting bed about to become a bountiful field of crops. I see millions of newly exposed weed seeds sitting atop a nice open seedbed, soaking up the sun, and properly prepped to become an invasive plant tangle. Instead, a no-till drill seeder, which creates and deposits seed in a shallow furrow and minimizes weed-producing soil disturbance, was to be our planting method.

Finally, how could I possibly have justified planting in July? My answer came with a word to which we keep returning: competition. Most native grasses are

This Pennsylvania meadow is on the rich soils of an alluvial floodplain. While it is a successful meadow, the plant diversity is not as great, as the high levels of fertility allow the aggressive species to grow so tall and dense that the more diminutive species are outcompeted.

top The soil on the large meadow site in Salisbury, Connecticut, was not amended in any way, as increased soil nutrients would favor weedy growth. The soil was also not tilled, which would have provided opportunities for weed seeds in the seed bank to germinate.

bottom The planted meadow is highly diverse, not in spite of but because of the poor soil conditions.

Scarlet Indian paintbrush, pinnate prairie coneflower, and nodding onion grow in the heat of summer.

Fields of rattlesnake master and tall blazing star and aster species in a mixed composition of blackeyed Susan and native grasses provide extensive habitat patches for local insects and birds.

A researcher from the Hawthorne Valley Farmscape Ecology Program conducts a study comparing insect diversity in designed meadows and agricultural fields.

heat-adapted species that do not actively grow until summer. Many associated flower species, such as pinnate prairie coneflower (*Ratibida pinnata*), scarlet Indian paintbrush (*Castilleja coccinea*), rattlesnake master (*Eryngium yuccifolium*), and nodding onion (*Allium cernuum*), are heat adapted as well. The most problematic weeds grow mostly in spring. Planting at the cusp of summer, when the native grasses are at the beginning of their growth period and the weeds are at the end of theirs, gives the meadow plants a head start and a competitive advantage.

So immediately after the July Fourth weekend, we planted the meadow. It turned out well. I wouldn't be telling you this story if it hadn't. Certainly, the folks who have wandered the paths these many years always seem to find inspiration. Possibly the most telling observation is how and by whom it is being used. One of the owners starts most mornings walking those paths, and a survey by local researchers revealed the presence of a large number of insect species.

Long after the meadow had been established, I met with a landscape contractor who was scheduled to work on the property. He was a knowledgeable horticulturist and the second-generation owner of a highly respected local firm. It was his first visit to the property, so I showed him the meadows and explained what we had and hadn't done to develop them. I explained how the plants, far from being babied, had received no inputs their entire lives; how new plants entered, dropped out, or relocated according to their own life spans and life cycles; and how derivative native meadow patches were now appearing, unplanted, in sunny woodland gaps elsewhere on the property.

We stopped along the path to look at a particularly nice composition. The flowers of broadleaf ironweed (*Vernonia glauca*) and Indiangrass had formed a beige and blue tapestry. On the ground layer, a clump of shiny green sweet fern had expanded into the cluster, and was mingling intermittently with the tall, upright stems. The ironweed had been planted live. The Indiangrass had been planted elsewhere from seed and migrated to this spot. The sweet fern had colonized entirely on its own accord.

He turned to me and said, "This is how plants should be grown."

After fourteen years, this meadow is beginning to take on the intricate character of a mature prairie.
In eastern North America, this level of intricacy is difficult to achieve in a highly fertile soil.

CREATING MEADOWS AND PRAIRIES

Native meadow is likely the best answer to North America's overreliance on lawn, a means of saving the enormous quantities of water, fertilizer, herbicides, pesticides, and fossil fuels annually invested in the cultivation of turf. Unlike shrublands and forests, meadows can be established from seed in a relatively short time and consequently are easier to install and less costly. In addition, people seem to gravitate toward open space and expansive views, attributes not common in a forest or shrubland.

While meadow is not a universal replacement for mowed lawn—it cannot compete with clipped turf, for example, as a play space for sports—it can substitute as a groundcover where these activities are not intended. And it is by far a superior treatment from the perspectives of maintenance requirements and wildlife habitat. Additionally, few would argue that a monotonous carpet of green can compete aesthetically with a well-grown meadow in full bloom.

Why then, even in an age of growing ecological awareness and appreciation of nature, do lawns continue to dominate our landscapes? Certainly meadows have been on the radar of most gardeners in the United States ever since Lady Bird Johnson advocated the planting of wildflowers in the 1960s. I believe the main reasons are the way in which meadows have been misrepresented to the public and the poor quality of most installations.

Advertisements associated with early meadow seed mixes represented the inevitable result of planting as a sheet of blazing floral color, suggesting that once established this amazing display would become permanent and entirely self-sufficient. If this sounds too good to be true, that's because it is. In fact, these mixes were, and still are, mostly composed of annuals, biennials, and short-lived perennials. These sorts of plants are not robust enough to compete successfully with the more aggressive weeds. If you put showy but wispy flax, poppies, and cosmos in the ring with heavyweights like Canada thistle, Japanese honeysuckle, and Oriental bittersweet, it's not hard to guess who will be left standing after a round or two.

Many of these short-lived meadow offerings rely heavily on plants native to arid climates, including the American Southwest. In such locales, vegetation is sparse and seeds

Where limestone bedrock creates high pH soils, like on this Connecticut property, Indian paintbrush thrives.

from short-lived mother plants can perpetuate the species by germinating in the open soil between the scattered plants. Therefore, mixes based primarily on short-lived, albeit native, species may have legs in arid regions. In most of North America, however, greater rainfall results in dense ground-layer vegetation, a condition where competition prevents seed germination. Consequently, the commonly offered claims that these plants will readily reseed in perpetuity wherever they are planted is patently false.

Why is it so difficult to find a seed mix that provides the plants needed for the long haul? The answer can be found in an examination of the different reproductive characteristics of long- and short-lived herbaceous plants. With their short life span, annuals, biennials, and short-lived perennials depend on seed production for their long-term proliferation and consequently produce large numbers of flowers and seeds early in life. As a result, their seed is cheap to produce and their floral impact is immediate. Long-lived perennials, on the other hand, that plan to stick around through good times and bad are more interested in expending their initial energy on root growth. Flower and seed production is a low priority at this early stage of their life cycle. The disadvantage of this, from the gardener's perspective, is that a seed mix based on these long-term survivors takes several years to produce flowers and will inevitably cost more. The advantage, of course, is that you'll have something to show for your hard-earned money beyond the first year or two.

Even the all-perennial mixes sometimes offered for sale do not usually include enough long-term species to make the planting competitive beyond the early stages. In other words, there are perennials and there are *perennials*.

In addition, rarely are the species selected for their suitability for a specific region. Even a mix that is designated as regional is not adequately targeted. Is a Rocky Mountain mix appropriate for both a home in the foothills and a cabin on a high-elevation ski slope? Will the same New England mix do equally as well in a moist floodplain as on a rocky hillside? Obviously not.

The early mass-marketed mixes often did yield brilliant color in the first summer. But the annuals died at the end of the first growing season, weeds took over in the second, and the frustrated homeowners threw up their hands and returned to mowing by the third. Understandably their attitude toward meadows veered from unrealistically optimistic to unnecessarily pessimistic, and they concluded that "meadows just don't work." This is not surprising. If you make it all about the flowers, eventually it will become all about the weeds. Possibly the worst part of this fiasco was that it squandered a great deal of good will toward these plantings and set back the effort to create viable native-oriented meadows by many years.

Meadows can work, but only if the gardener selects plants and uses techniques that reflect the local habitats and ecological processes that will affect their survival and proliferation. This is why a successful meadow will look quite different in different regions.

Travel the central Texas highways in February or March, and your views will be replete with waves of bluebonnets (*Lupinus texensis*) and Indian paintbrush. Arrive at the Lady Bird Johnson Wildflower Center in Austin and you will see wonderful planted examples of the shortgrass prairie of the Texas hill country. Not coincidentally, these will include the same bluebonnet and Indian paintbrush species that you saw in the wild.

Light patterns, sky, and the interplay of angular lines all contribute to the visual character of this 40-acre meadow in Salisbury, Connecticut.

Wander the paths of the Morton Arboretum's Schulenberg Prairie planting outside of Chicago and you'll experience a wonderful recreation of the tallgrass prairie that once blanketed much of the upper Midwest. This incredible planting, which was conceived and executed by botanist Ray Schulenberg beginning in 1962, spans more than 100 acres. Schulenberg and his associates collected seeds from wild populations in the Chicago area, grew small plants in the greenhouse, and directed an army of volunteers, who planted the prairie over a ten-year period. Walk the paths and peer into the grasses and flowers, and you will observe how beautifully complex and interrelated a true prairie composition actually is. The meaning of plant community will become viscerally apparent.

A visit to the Crosby Arboretum in Picayune, Mississippi, will allow you to experience a different type of meadow. The yellow pitcherplant (*Sarracenia alata*) bog here may be one of the most exotic appearing landscapes you will ever visit, except it is entirely native. And not just native to the state, but to the specific soil, sunlight, and hydrologic conditions that exist on that exact spot in Mississippi. That's why it's still there today and will likely be there for years to come.

Mt. Cuba Center for the study of Piedmont flora in Hockessin, Delaware, sits in the horticultural mecca defined by the grandiose plantings of the Winterthur Estate and Longwood Gardens. Yet even these garden giants won't diminish your appreciation of Mt. Cuba's unassuming pocket meadow in autumn, where buff-colored native grasses add a quiet contrast to the blazing orange, red, and yellow foliage of the surrounding forest.

In order to satisfy the goals of long-term sustainability and low maintenance, you must design the meadow as a functional plant community first and a flower garden second. Only by understanding and incorporating the compositions, patterns, and processes inherent in naturally occurring grasslands can we create landscapes that will be viable over long periods of time without massive amounts of assistance.

Does this mean that we must sacrifice aesthetics for sustainability? Not at all. True, you won't see the summer-long wall-to-wall carpet of colors commonly portrayed on wildflower seed packets, but, if you do your work right, you'll see something even more beautiful and far more complex and subtle. Similar to the contrast between an annual and a perennial border, the flowers of the perennial meadow will come in waves, with the bloom shifting from one area to another as the season progresses. Native grasses glimmering in the sun and swaying in the breeze, complemented by sweeping drifts of wildflowers, will provide a truly spectacular scene. What's more, rather than waiting in dread to see what will emerge the second year after the first year's annual flowers have expired, you can enjoy an ever-evolving natural landscape that changes from season to season and year to year, taking its own unique and continually beautiful course.

What of the carefree aspect that the instant meadow purveyors often claim for their product? There is truth to this, but only in that it doesn't take much effort to maintain something that disappears. Don't believe it if someone tells you that a native perennial meadow is carefree. Nothing is. I once spoke to a gentleman who had just spent two hundred dollars steam-cleaning his AstroTurf back yard.

In the first year or so, a meadow planting can take significant effort, particularly on a weedy site, but once the meadow is established by the third or fourth year, the dense

Bluebonnet grows in a natural area at the Lady Bird Johnson Wildflower Research Center in Austin.

top Yellow pitcherplants dot
a savannah at the Crosby
Arboretum in Mississippi.

right Little bluestem provides
a tawny foil to autumn foliage
of deciduous trees at the Mt.
Cuba Center in Delaware.

plants do most of the weed suppression for you. Management inputs will drop dramatically—never to nothing, but certainly less demanding than virtually any other type of planting, whether lawn or ornamental garden. It should be understood, however, that the interventions it does demand require more mental agility than simply firing up the lawn mower.

The truth is, though, that wildflower meadows, if planned, installed, and managed properly, can contribute tremendously to naturalizing the North American landscape. When integrated into a well-designed landscape matrix, a meadow can help transform a residential property into a beautiful, wildlife-friendly, and stimulating home environment while vastly reducing the time spent with a noisy, polluting mower.

This is an admirable and desirable goal, but how do we achieve it? Despite the difficulties gardeners have too often experienced in the past, meadows can be consistently successful if you follow the proper procedures for design and implementation. In this chapter, I present the basic analytic thought processes, procedures, and equipment that can help you to guide a meadow into fruition, but there is always more than one way to skin a cat. The more you understand the underlying principles that cause these approaches to work, the more you will be able to adjust them to your own goals and the unique situations you will encounter. I once heard Jean Marie Hartman, an accomplished ecologist and landscape architect from Rutgers University, say that there is only one answer to every question about establishing wild landscapes: "It depends."

FURTHER SITE ANALYSIS

This analysis follows the procedure for more general site analysis laid out earlier, but includes as well several points more specific to meadows.

THE IMPORTANCE OF LIGHT

The first characteristic to look for is light exposure. Full sun for at least six hours a day is a requirement for a meadow planting. Insufficient sunlight will favor woody species over herbaceous wildflowers and grasses, causing an increase in maintenance requirements. By undercutting the meadow plants, it will also favor weed invasion.

This may be a small meadow, but it's still too big to hand weed like a garden. Consequently, habitats and ecological processes must be considered when designing, planting, and managing it.

KNOW YOUR SOILS

Soil type is the next consideration. Meadows can thrive on a variety of soil types, including sand, loam, or clay, but each one requires a different suite of adapted plants. For this reason, a soil test is an absolute necessity. The goal of testing, however, is different from that of the traditional gardener. Someone cultivating a meadow wants to determine not what amendments are needed to create the optimal growing condition for all plants, but to determine which plants will thrive in the existing condition without any amendments.

As a rule, I advise against amending the soil. A so-called bad soil, in particular one that is poorly drained or very dry, can provide a competitive advantage to the meadow species (assuming you selected those species that naturally occur in this condition). Many weeds favor richer soils and compete less successfully on poor ones against the native flowers and grasses that were selected for adaptation to the site. Victory for the meadow plants lies in how successful you are in selecting suitable species, even on poor soils. For example, white wild indigo, butterflyweed and flowering spurge (*Euphorbia corollata*), all highly ornamental plants, grow well on dry, sandy soils, whereas spotted joe pye weed (*Eupatorium maculatum*), pink turtlehead, and New England aster (*Symphyotrichum novae-angliae*) thrive on poorly drained clay soils. Of course, some weed species are adapted to the existing soil on your site, even if it is a poor one. But many weeds won't be, meaning at the very least you will be dealing with fewer problematic species than if you create a fertile condition where every weed in the area is more than happy to come knocking on the meadow's door.

In general, then, the nutrient levels in your soil are far less important for a meadow planting than for traditional horticultural features. Barring an extreme deficiency, you should not fertilize as this will typically foster the weeds more than the meadow species. On the contrary, when planting a meadow on rich soils an application of sulfur, which lowers the pH of the soil, can help ease weed pressure. This ties up nutrients that are normally available to plants at higher pH, essentially mimicking infertility. This application may only be necessary for the first few years after planting. Once the planted species have become dominant, weed pressure will diminish to competition from the meadow. Alternatively, you can plant a mix of more aggressive species to combat the weed onslaught that will surely come when you disturb a rich soil. Mountainmint species (*Pycnanthemum* species), bergamot (*Monarda* species), and many native grasses can fulfill this role.

GRADE AND TOPOGRAPHY

A meadow design can also be affected by grade and topography. A north-facing slope may not be as favorable to meadow plants, especially in northern regions, as it receives less direct sunlight. Because this condition generally favors woody plants, shrubland or reforestation may be a better choice in this situation. If a meadow is highly desirable for aesthetic or practical reasons, the area can be treated similarly to the overly fertile condition described above, where early-stage sulfur applications or highly aggressive native meadow species are used to combat the expected weed pressure.

If you intend to plant a meadow in a low-lying area, one that remains wet through a rainy spring, you must select plants adapted to that condition. You should also take note of variations or microhabitats within the site, such as patches of dry, thin soil, and modify the planting choices there to suit.

Topography can also affect the planting schedule. Spring, early summer, or late autumn are all suitable times to seed meadows, but a sloping site may require a spring seeding, because autumn-planted meadow seeds typically remain dormant until spring and are liable to wash away on a slope over the course of a winter.

EXISTING VEGETATION

Analyzing existing growth on and adjacent to the site is especially important with meadow plantings, for it yields extremely valuable information regarding what plants will grow well on the site and which weedy species are likely to be a problem. If a native meadow species already occurs on the site, include it in your meadow palette. If it occurs there naturally, it is adapted. If, on the other hand, a problematic weed grows on or near the site, you should eradicate it before planting to avoid future infestations.

dominated by European cool-season grasses in North America rarely contain significant numbers of American native wildflowers, but they do often contain wildflowers of similar European origin, such as Queen Anne's lace (*Daucus carota*). Plants in a community fill spatial niches in a complementary manner, both above- and belowground. European cool-season grasses have mat-forming root systems, and these mats are dense but shallow. Queen Anne's lace is in the carrot family, and in common with many European field species, it has a deep taproot. So the root systems of these two plants are operating in completely different soil zones and not directly competing for water and nutrients. In other words, they coevolved to occupy separate but complementary spatial niches, in this case belowground. Consequently, they cohabitate quite easily, even here in North America.

Conversely, native North American meadow wildflowers largely coevolved with deep-rooted native clump grasses, with space existing between the grass roots for wildflowers with various root depths. One native wildflower that you do occasionally find in cool-season European grass pastures in North America is butterflyweed. I suspect it is not a coincidence that this plant has a deep taproot. So the lesson when it comes to grasses is don't mix and match. And if you do want to plant native wildflowers into a meadow with grasses of European origin, consider niche in your selection.

While you won't be surprised that my preference lies with native over exotic meadows, meadows of cool-season European grasses are certainly more ecologically sound and easier to manage than turf, and they may be a good choice in some instances.

Cool-season grasses

Pros Quick to establish, attractive spring foliage, wildlife value (primarily from cover), already exist in most grass-dominated fields in North America, less expensive seed.

Cons Support less wildflower diversity, require additional mowing during the growing season to avoid brown matted-down appearance in summer.

Warm-season grasses

Pros High potential for wildflower diversity, greater benefit to pollinators (many of which are specialists to particular wildflowers), provide crucial habitat for declining grassland bird species, attractive in summer, autumn, and winter, require one mowing per year once established.

Cons Slower to establish, more expensive seed, dormant in spring.

This planted patch of Indian paintbrush is continually expanding. The fact that this species occurs naturally on limestone soils and is planted in a meadow with limestone soils likely has much to do with its ability not only to survive but to proliferate.

ASSEMBLING A PLANT COMMUNITY

Given that European grass meadows are by the far the most common occupiers of open fields in North America, and given the paucity of native ones, I will spend the remainder of the chapter discussing native meadow establishment.

I've already discussed the importance of filling niches. In the case of meadows, there are some special considerations. Once again your goal should be to mimic the growth of natural grassland. Even a cursory look at a mature prairie or wild meadow will reveal an incredibly dense tapestry with a canopy, numerous middle layers of growth, and a creeping understory. This dense interweaving of stems and foliage should be mirrored underground by the root systems of the meadow grasses and wildflowers, with shallow, spreading mats of roots that fill the upper layer of soil complemented by deep-reaching tap and fibrous roots that fill the soil underneath. Such a dense fabric of growth, you will find, is remarkably weed resistant once established. To give a better idea of that density, consider that I plan my seed mixes to spread 80 to 175 seeds per square foot of soil. That sounds like a lot, but most of the seeds won't germinate or survive the competition if they do. (It's a tough world out there.)

I've also mentioned the need to fill niches in time. There are two time scales to consider with meadows. The first is seasonal. Plants such as the native meadow grasses grow most actively during the warm weather season, from late spring until early autumn, whereas others make their growth during cool seasons, especially spring. To neglect either group in your planting is to invite an invasion of weeds with a corresponding season of growth.

The second time scale is one of years. To keep the weeds at bay, a meadow should include fast-growing plants that cover the ground during the first year of growth, biennials and short-lived perennials to take over as the first year plants fade, and long-lived perennials to provide long-term cover. All of these need to be present to prevent a vulnerable gap in the meadow's ability to resist weeds.

There is one additional niche to fill in the meadow, one that is neither spatial nor temporal. This is the ecological function of nitrogen fixation. Plants in the legume family fix nitrogen from the atmosphere and add it to the soil, which produces a modest increase in fertility. Whereas the high levels of fertility that fertilizers produce pose a threat to meadow plantings, the level of fertilization furnished by legumes is typically perfectly suited to the meadow plants that coevolved with the nitrogen-fixing plants. Plant decomposition also affects fertility. Again, if you're using a plant community model, the fertility levels that result from the decomposition of those plants should be optimal to support that community.

The chart provided here lists the various spatial, temporal, and fertility niches needed in an ecologically sound meadow planting, with examples of some plants that fill those niches. When assembling the plant community for your meadow planting, be sure that each of these niches is represented by several species. If a niche has no or few plants listed, select additional species to fill that niche.

MEADOW NICHES

Sample Species Organization Chart

Aboveground space

Upper level

Baptisia australis blue wild indigo

Echinacea pallida pale purple coneflower

Monarda fistulosa wild bergamot

Middle level

Asclepias tuberosa butterflyweed

Solidago speciosa showy goldenrod

Symphyotrichum laeve smooth aster

Ground level

Bouteloua curtipendula sideoats grama

Dalea purpurea purple prairie clover

Liatris aspera tall blazing star

Belowground space

Surface + rhizomatous

Liatris aspera tall blazing star

Monarda fistulosa wild bergamot

Rudbeckia hirta blackeyed Susan

Middle + deep fibrous roots

Bouteloua curtipendula sideoats grama

Dalea purpurea purple prairie clover

Symphyotrichum laeve smooth aster

Deep taproot

Asclepias tuberosa butterflyweed

Baptisia australis blue wild indigo

Echinacea pallida pale purple coneflower

Seasonal time

Cool-season species

Hordeum jubatum foxtail barley

Monarda fistulosa wild bergamot

Tradescantia ohiensis bluejacket

Warm-season species

Asclepias tuberosa butterflyweed

Bouteloua curtipendula sideoats grama

Liatris aspera tall blazing star

Time in years

Initial: year 1

Agastache foeniculum blue giant hyssop

Elymus canadensis Canada wildrye

Rudbeckia hirta blackeyed Susan

Midterm: years 2–4

Echinacea pallida pale purple coneflower

Monarda fistulosa wild bergamot

Tradescantia ohiensis bluejacket

Long term: years 5+

Asclepias tuberosa butterflyweed

Baptisia australis blue wild indigo

Bouteloua curtipendula sideoats grama

Nitrogen-fixing legumes

Baptisia australis blue wild indigo

Dalea purpurea purple prairie clover

Thermopsis villosa Carolina bushpea

Wildflowers and grasses in this field were all seeded at the same time. The lanceleaf tickseed is a short-lived perennial that can dominate the meadow early on, and it will be supplanted in subsequent years by more slowly growing long-lived species like wild quinine and a variety of native warm-season grasses.

As the meadow matures, long-term players such as blue wild indigo and Culver's root can take five to ten years to have a visible presence if grown from seed. If their presence is desired earlier, they can be added in as container-grown plants.

top In this composition, the sunflower dominates with the coneflower remaining, but in a lesser role. The little bluestem and butterflyweed are outcompeted.

bottom left Which are the long-term players in this meadow composition? The biennial blackeyed Susan will phase out. The coneflower and oxeye sunflower will persist. As the most aggressive species in the composition, the sunflower will likely be the dominant player. The blackeyed Susan may reappear from seed in the event of disturbance. If the site is dry, the butterflyweed and little bluestem will remain competitive and persist. In fertile conditions, these species will likely be outcompeted by the taller species, particularly where water and nutrients are easily available and allow them to achieve their full stature.

bottom right The dry, infertile conditions on this south-facing slope limit the growth of more aggressive meadow species, allowing the little bluestem and butterflyweed to persist longer.

This all brings to mind a Frank Egler comment I quoted earlier, "Nature is not more complex than we think, but more complex than we can think." Given that we are trying to understand natural processes and piggyback on them where possible, that complexity may seem discouraging. But it need not be if you are assembling plants according to community dynamics. Consider the following overall process:

- Determine the best adapted plant community for your habitat.

- Identify as many of the processes and interactions that govern that community as you can.

- Put a long-term program in place that includes these processes and interactions.

- Let nature work out the rest.

- Finally, observe how it all plays out and learn from it.

SELECTING FOR AESTHETICS

Placing plants according to their environmental preferences within the meadow not only increases their chances of survival, it also provides the inherent visual patterns that define that meadow's overall aesthetic character. Flowers, seed heads, stems, and foliage provide the details. Ending it there can result in a perfectly beautiful meadow, but there are plant selection decisions that can allow you to tailor the look of the meadow to suit your personal taste. Thus, you can determine combinations of flower colors and forms, foliage textures, and succession of bloom as you would in any well-planned garden. In addition

below left Purpletop tridens, a native bunchgrass, forms drifts across this Arkansas field. These patterns, which are likely derived from different moisture levels in the field, contribute to the beauty of the composition.

below right A similar pattern was intentionally designed into this planted meadow.

This turf will be replaced by a small meadow garden.

In the second year of growth, short-term species like blackeyed Susan and foxtail barley are visually dominant. Long-term, slow-growing species are present but not yet flowering, dedicating their resources instead to root development.

The relay has occurred. The blackeyed Susan and foxtail barley have yielded to purple coneflower and mountainmint. All were planted at the same time, but developed at different rates. Both short- and long-term species were needed for the planting to outcompete weeds at all stages of meadow development.

to accommodating environmental preferences, drifts can be placed purely for aesthetic effect and can be arranged with the same artistic subtlety that any creative designer would employ.

SELECTING FOR WILDLIFE HABIT

I guarantee that simply by planting a native meadow you will create significant wildlife habitat. In fact, you will create the most endangered type of habitat in North America. Every year millions of acres of abandoned farm fields are replaced by development or revert to forest. The stems, seed heads, flowers, and overall cover that will be provided by the grasses and wildflowers such as blazing star and coneflower (*Rudbeckia* species) will attract insects and birds to the meadow. Meadow plants like bergamot and beardtongue (*Penstemon* species) are particularly attractive to hummingbirds, and seeds of many meadow species are eaten even in wintertime.

Of course, you can go beyond this and select plants to accommodate uncommon species with very precise requirements. Obviously, significant additional research would be required to take it to this level. In some cases, the planting could include woody species that would grow not inside but in the vicinity of the meadow. The prothonotary warbler nests in wooded wetlands, but any wet woodland is not good enough. It uses, among other things, the bark of river birch (*Betula nigra*) or bald cypress (*Taxodium distichum*) to line those nests. If your site has a wet area, plant either of these trees if they are not already present and you will have completed the picture.

FORMULATING A SEED MIX

It's now time to convert the plant list you assembled for each meadow area into a seed mix. This largely consists of quantifying the individual species, and the first step is to assign a relative percentage to each. This can be based on a number of factors. A fairly diminutive species may require a higher percentage in order to have a visual presence. Conversely, you may want to limit an overly aggressive species to a low percentage to prevent it from dominating; this can also be achieved by keeping aggressive species out of the main mix and spot sowing them in isolated patches.

Budget can also be a consideration. There are vast cost differences from species to species, and you may want to limit or exclude those that will bust your budget. More than once I have been shocked by the cost of a seed mix only to find upon closer examination that as much as half of that cost was tied up in one species. Unfortunately, some pretty good plants, including many of the milkweed species, fall into this category. If you can't afford much of such an expensive seed but want to give it a visual presence in the meadow, sow it in patches where it will be a prominent part of the views, rather than diluting its impact by distributing it throughout. In this fashion, a plant that contributes a small percentage to the overall mix can have a disproportionate impact.

Once you have assigned relative percentages, you need to convert them into weights. Although this is the measure that you'll ultimately order the seed by, the number of seeds per species is the measure that should determine the relative abundance of each. Let's look at two plants to illustrate why. Blue wild indigo produces about 1500 seeds per ounce, whereas joe pye weed (*Eutrochium fistulosum*) produces approximately 95,000 per ounce. So, if you plant one ounce of each you will be planting more than sixty times as much joe pye weed as blue wild indigo, and you'll end up with a joe pye weed meadow.

In addition to perennial grasses and wildflowers (forbs), you'll also want to include a nurse or cover crop in your seed mix: a quickly germinating, clump-forming annual grass, such as oats or intermediate ryegrass. By providing quick coverage of the site, the nurse crop helps to reduce weed invasion and soil erosion during the first season after planting. This is very important because the meadow is most vulnerable at this time, as the longer lived perennials and grasses are not yet well enough established to stabilize the soil. These annual grasses are commonly used for this purpose in the construction trades. For meadow use, however, it's important to cut the recommended rate of seeding at least in half, which will allow the nurse crop to stabilize the erodible soil without outcompeting the newly planted seeds.

Incorporate the nurse crop into the seed mix, and you have a custom-designed seed blend that contains species names and quantities for each (expressed in weight), which you can now submit to a seed house.

QUANTIFYING THE SEED MIXTURE

A = square footage of the project area

B = total number of seeds per square foot (typically 80 to 175)

C = percentage of the mix of each species

D = number of seeds needed for each species = $A \times B \times C$

For example, let's consider the case of planting a 1000-square-foot area to be seeded at 90 seeds per square foot with a mixture of 70 percent joe pye weed and 30 percent blue wild indigo. For joe pye weed:

D = 1000 square feet × 90 seeds/square foot × 0.70 = 63,000 seeds of joe pye weed

For blue wild indigo:

D = 1000 square feet × 90 seeds/square foot × 0.30 = 27,000 seeds of blue wild indigo

To convert the seed number to weight of seed, divide by the number of seeds per ounce (which can be obtained from seed house catalogs or online searches). For joe pye weed:

63,000 seeds ÷ 95,000 seeds/ounce = 0.7 ounce of joe pye weed seed

For blue wild indigo:

27,000 seeds ÷ 1500 seeds/ounce = 18 ounces of blue wild indigo seed

If this bit of mathematics seems intimidating, there is a work-around. Figure out the values of A, B, and C and submit those figures to a reputable seed house, which can calculate the weight of seed that you need for each species. Just make sure to let the seed company know that the percentages you are submitting refer to seed counts, not weight.

This seed mix for an upland meadow site may seem to contain a lot of species, but approximately one-third are early-succession species that will drop out as long-lived plants take over.

Seed Mix Specifications
Project: Iron Mountain House
Mix: Main Mesic
Site Location: CT

Prepared by:
Larry Weaner Landscape Design

Area in acres = 1.8
Area in sq. ft. = 77828
Total seeds per sq. ft. = 125
Percent grasses = 60%

Results

24.5 lb. seed/acre (excl. nurse)
43.26 total lb. this mix (excl. nurse)
Total spp. (excl. nurse) 25
Number of graminoids 6
Number of forbs 19

Botanical Name	Common Name	Qty	Cost	Seeding notes	Substitutions/ comments
Grasses, sedges, and rushes					
Bouteloua curtipendula	sideoats grama	97.29 oz.			
Carex conoidea	openfield sedge	11.67 oz.			
Carex vulpinoidea	fox sedge	5.84 oz.			
Elymus virginicus	Virginia wildrye	138.98 oz.			
Schizachyrium scoparium	little bluestem	175.11 oz.			
Tridens flavus	purpletop tridens	41.25 oz.			
Forbs					
Allium cernuum	nodding onion	15.36 oz.			
Asclepias tuberosa	butterflyweed	18.10 oz.			
Cassia fasciculata	partridge pea	5.66 oz.			
Castilleja coccinea	scarlet Indian paintbrush	2.95 oz.			
Coreopsis lanceolata	tickseed	43.24 oz.			
Echinacea purpurea	eastern purple coneflower	1.04 oz.			
Eryngium yuccifolium	rattlesnake master	9.73 oz.			
Lespedeza capitata	roundhead bushclover	41.27 oz.			
Liatris spicata	dense blazing star	20.75 oz.			
Monarda punctata	spotted beebalm	4.86 oz.			
Penstemon digitalis	foxglove beardtongue	21.23 oz.			
Pycnanthemum incanum	hoary mountainmint	2.16 oz.			
Rudbeckia hirta	blackeyed Susan	4.19 oz.			
Solidago nemoralis	gray goldenrod	1.42 oz.			
Solidago speciosa	showy goldenrod	2.11 oz.			
Symphyotrichum laeve	smooth aster	0.65 oz.			
Symphyotrichum novae-angliae	New England aster	2.05 oz.			
Tradescantia ohiensis	bluejacket	14.59 oz.			
Zizia aurea	golden alexanders	10.61 oz.			
Nurse crop					
Avena sativa	common oat	35.38 lb.			

Total cost

Notes:
1. Use the following eco types if available:
Northeast
PA, NY, NJ, CT, MD

PURCHASING THE SEED

It is best to submit your seed mix to multiple seed houses. In addition to shopping for the best price, one seed house may not carry all of the species in your mix. Request that they package the species individually, allowing you to mix the companion species once they are all received. As will be explained shortly, it may be necessary to separate seeds by size and consistency (small, medium, and large fluffy) depending on the planting method you need to accommodate. This obviously can't be done if you have received the seeds pre-mixed. Actually, opening the various seed packages can be quite interesting. My clients are always fascinated to peer into a bag of tiny, dust-like sedge seeds (don't sneeze), followed by a look at the fat bean-like seeds of Carolina bushpea (*Thermopsis villosa*). The seed of little bluestem, my favorite grass, feels like air in your hand as the majority of the bulk is near weightless fluff.

Legumes are sometimes packaged together. Before legume seed is combined with other large seeds, it should be moistened and mixed with an inoculant powder provided by the seed house. These powders contain beneficial bacteria that speed up the legumes' nitrogen-fixing ability.

I also recommend that you buy from seed houses that provide pure live seed. Not all of the seeds that a plant produces are viable, and the percentage that is viable varies from species to species. Viability levels also vary for the same species from year to year, usually related to the previous year's climate. The best seed houses test each species, each year, and record the percentage that is viable. If you need 1 ounce of cardinalflower seed, and its tested viability for that year was 50 percent, a reputable seed house will send you 2 ounces so you will receive 1 ounce of pure live seed. Another supplier may sell you 1 ounce of seed for 25 percent less, but if it is not pure live seed you will actually be paying more.

The desire to purchase regional ecotypes may also factor into where you buy your seed. We have already discussed how the same species may have genetic differences based on the region in which the plants evolved, so a regional ecotype is likely to be best adapted to that region. It doesn't necessarily matter where you bought the seed, but it does matter where the plants that provided the seed originally reside. This is more easily accomplished for midwestern plantings, a region where there are far more suppliers than in the East. Some seed houses do collect from various regions and keep records of that information. So get seed that originates from your region when you can. When you can't, you must decide whether to alter your list or settle for a source outside of your region.

USING LIVE PLANTS

For small meadow plantings, starting with container-grown plants rather than seed may be economically feasible. This method has the advantage of faster establishment, with the meadow maturing in a year or two rather than the multiyear time frame that starting from seed typically requires. In addition, you can obtain a more precise arrangement, allowing for a more gardenesque appearance.

You can also use a mix of seed and container-grown plants. Some meadow species can take five to ten years to reach flowering size, and these would be good candidates to plant live. All the wild indigos, Culver's root, and wild quinine (*Parthenium integrifolium*) fall into this category. And some plants don't become established when seed is sown directly into the landscape. These include Canadian anemone, meadowsweet (*Filipendula ulmaria*),

Carolina bushpea was installed as live plants in this birch grove at the edge of a larger meadow. Carolina bushpea is a legume that fixes nitrogen, thereby helping to provide this critical nutrient to other plants in the community. Including it here was not purely functional, however, as the plant's spike-like racemes visually echo the columnar white birch trunks.

and milkvetch (*Astragalus* species). If you want these plants in your meadow, you have no choice but to install them as live specimens.

Live meadow plants are packaged several ways. Typical perennial containers are fine, but using them can get quite expensive, even for very small meadows. In more recent years, smaller plugs have become available at a much lower cost. These are packaged in trays, similar to evergreen groundcovers like pachysandra or vinca. Look for "deep root" plugs, which are a modification that allows for higher survivability in the rough-and-tumble environment of the meadow. Bare root perennials are also an option, but I have generally found them to be impractical because they can be planted only while dormant.

PLANTING THE MEADOW

Site preparation prior to installing the meadow is of the utmost importance to achieving a successful planting. It begins with the elimination of any existing growth, as seeds will not grow if placed in competition with already established vegetation, including turf. Of course, you should survey the field first and mark the location of any existing desirable vegetation to avoid its elimination as well. The most common methods for this are repeated applications of short-lived herbicide sprays, repeated tilling, or a combination of the two. Tilling will bring to the surface dormant weed seeds that must be allowed to germinate and then shallowly cultivated or sprayed with herbicide before you plant the meadow.

Fewer weed seeds will be activated if you use no-till seeding techniques, such as surface raking or a no-till drill seeder. Both will prepare a seedbed without bringing dormant weed seeds to the surface and encouraging their germination. Surface raking is best for small areas and can be done by hand or with a walk-behind power rake. For large areas of several acres, the best planting tool is a native grass drill seeder. This tool is designed for use with

native grasses and wildflowers, whose seed vary so considerably in size and texture that typical agricultural seed drills cannot handle them.

Native meadow seeders have separate boxes for small, medium, and large fluffy seeds. Each box is designed to evenly deliver the seed to the drill, which then deposits it into a shallow furrow in the soil. This is why I suggested ordering seeds to be packaged individually by species, as it will allow you to group the species by size and place them in the appropriate box. Planting depth should be set at no deeper than ¼ inch to avoid burying the seed too deeply, a mistake that can have dire consequences for germination.

Although the no-till seeder looks like a bulky beast, it is performing a very precise procedure. Between the planted furrows there is no disturbance, and consequently less weed seed activation. Within the furrows there is disturbance, and the meadow seeds are the beneficiaries. This is targeted disturbance, and another example of an ecological principle informing practical methodology.

If you have used a no-till drill seeder, the seed is now planted. If you have tilled or surface raked, you now have a prepared seedbed and are ready to sow. Because meadow seeds are generally small, very little seed covers very large areas, and you'll find them easier to handle and distribute evenly if you bulk them up by blending them with some inert material. Sawdust or wood shavings work well. Moisten the bulking material slightly before mixing to ensure that the seeds adhere to it and so blend throughout evenly when the seeds and the bulking material are mixed. Spread the seed evenly over the area and use a rake to incorporate it about ¼ inch below the surface of the soil, tamp or roll for good seed-to-soil contact, and afterward mulch with clean straw.

The timing for seeding a meadow differs somewhat from that for seeding turf. Both can be planted in early spring, but meadow seeding can continue after the weather has warmed and even extend into early summer. Most North American prairie plants love heat, whereas many weeds, a large number of which are of agricultural origin, do not. Because of this difference, early summer may actually be the best time to plant a very weedy site. Sowing in autumn is limited to late dormant seeding after the weather has cooled. Because germination will not occur until spring, late autumn sowing is not recommended on sloped sites where erosion can be a problem.

If you are using seed and live plants in combination, you should defer the planting of the live plants until a full growing season after the sowing of the seed. As will be explained further in the management section of this chapter, monthly mowing over the top of the small emerging seedlings is used as a tool for weed control during the meadow's first growing season; mowing would harm the container-grown plants, which are much larger than the seedlings at this stage. My team often seeds in spring, mows for the first summer, and installs the live plants in autumn. An added advantage of autumn planting is that it allows the plants two full growing seasons (autumn and spring) before hitting their first hot summer, thus decreasing the need for watering during the establishment period.

What about watering the newly planted meadow? Of course, live plants cannot be allowed to dry out until they are reasonably well established. But seeds are just fine on their own, and they will wait for sufficient rains to germinate.

top A walk-behind dethatcher is being used to install a small meadow. Walk-behind equipment can also prove useful in larger meadow installation in areas inaccessible to tractor-pulled equipment.

middle Native grass seed comes with a fluffy chafe that can clump and clog the tubes of this native grass box. The box has metal agitators that keep the seed flowing evenly.

bottom The rollers press the seed into the furrow for good seed-to-soil contact.

top A native meadow no-till drill seeder with three boxes for various sized seed is being used to install a large meadow. The area to be seeded was treated with herbicide to eradicate existing growth and then cut closely to the ground. The seeder is cutting shallow furrows through the dead grass and depositing the seed.

middle The trash tills cut the furrow.

bottom The furrows after seeding.

top Many native wildflowers have very small seed. This box has a very fine calibration mechanism that can prevent this small seed from flowing out too quickly.

middle The drills drop the seed into the furrow.

bottom This seeded meadow exhibits the density and fine-grained diversity of a mature planting.

Grass or wildflower drifts can be added to a meadow purely for aesthetic effect. If you have selected plants for this purpose and haven't already located them on a plan, they can now be located by setting them out in the field. Look for special features in the landscape, like an ancient tree or a rock outcrop. Then place a drift that leads the eye directly to it. Tall drifts can also be located, as in any well-designed garden, to frame or slowly reveal views or special features. These compositions, if artfully placed, can add a great deal of character to the planting, even in the early stages when the meadow is still developing.

AN ALTERNATIVE APPROACH TO LAWN

Although widely regarded as an environmental affront, lawn can be a compatible component of an ecological garden, especially if it is treated as a short, mowed meadow.

The first imperative when planning lawn areas is to limit their size and to restrict this type of planting to areas that truly need the practical qualities of turf. Thus, an area intended for outdoor entertainment could benefit from the traffic tolerance and easy walkability of mowed grass, as could a spot designed for sitting and sun-bathing, but neither need be large. Above all, lawn should not become the default groundcover as it has been traditionally in North American domestic landscapes.

The next step in designing an ecologically compatible lawn area is to assess the soil in the proposed area. Have a sample tested for soil type, pH, and fertility, which this will help you determine which grasses will grow best and with the least intervention on your site. Carefully observe the drainage of the soil. Does it remain sodden for a day or more after every rainstorm? Or does the soil dry rapidly, tending to be drought-prone? Finally what is the exposure of the area, sunny or shady? The answer to these questions will help you select a grass species that is well-adapted to the site and that will flourish with fewer inputs of fertilizer and without pesticides.

To reduce the labor of maintenance and the lawn's impact on the environment, consider planting a grass that is naturally compact and rarely requires mowing, such as one of the fine-leaf fescues (Chewing's or red fescue, *Festuca rubra*, or hard fescue, *Festuca longifolia*) in the Northeast, buffalograss (*Buchloe dactyloides*) in the West, or centipede grass (*Eremochloa ophiuroides*) in the warmer parts of the Southeast. Sheep fescue (*Festuca ovina*) is particularly well adapted to poor, dry, and acidic soils in cooler regions. Your local cooperative extension agent can help you select the grass best adapted to your region and site.

Enhance the competitiveness of the grass by obtaining seeds inoculated with endophytic fungi. Tall fescues, perennial ryegrasses, and fine fescues all host these beneficial fungi that live within the grass plant, making it distasteful to leaf-eating pests and resistant to disease.

To increase the biodiversity of your lawn and make it more hospitable to pollinators, select a species that is bunch-forming rather than mat-forming and interplant the grass with low-growing, sun-tolerant wildflowers native to your region; violets (*Viola* species), wild strawberry, and azure bluets (*Houstonia caerulea*) are examples of flowers suitable for the Northeast and Mid-Atlantic region. These wildflowers are best inserted into the lawn as plants in the spring after the chosen grass has covered the area. These inserts will require some special care at first, in particular irrigation during dry spells, but if you have chosen species adapted to the site, they will soon root in and with time proliferate and spread throughout the lawn. As in a meadow, including adapted forbs such as these will help to fill ecological niches that would otherwise prove inviting to broadleaf weeds. White clover (*Trifolium repens*) serves a similar function and also enhances the fertility of the soil; however, it should not be used if a meadow is on or intended for the property, as

Don't be afraid to be bold. A mass of towering cup plant, with its cheery yellow flowers and coarse foliage, serves as a backdrop for grasses, blazing star, and other wildflowers. With rhizomatous roots, cup plant forms substantial colonies and so can be used to create a major visual statement in large meadows.

white clover can be a problematic weed in the early developmental stages. Various *Oxalis* species can be a good substitute.

Late summer or early autumn is the best time to sow a new lawn in northern regions where cool-season grasses are the norm, whereas spring planting is preferred for warm-season grasses in southern regions where warm-season grasses predominate. In either case, weed control will be a necessity as the new lawn establishes itself, but once the grass has established a dense cover, it should be naturally resistant to weed invasion. In fact, if you select a grass seed that is well adapted to your site and intermingle legumes and other forbs, maintenance should be minimal. Whatever type of grass you select, you should avoid the kind of frequent fertilization and liming recommended in lawn maintenance manuals.

MANAGING THE MEADOW

An understanding of ecological succession is extremely important for the maintenance of the meadow. This is the process by which a disturbed area progresses naturally from herbaceous meadow (first annuals and biennials, then perennials) to woody shrubs and pioneer trees and finally to a mature forest. In dry regions, various forms of prairie can be a stable and long-term composition. In establishing a permanent meadow where woods naturally predominate, we are arresting the process of ecological succession at the herbaceous perennial stage. By understanding nature's next move throughout the process, we are better able to make intelligent decisions resulting in fewer maintenance requirements and a more successful end result.

Although the meadow, once established, will require substantially less maintenance than mowed lawn, the first one to two years will require guidance in order to achieve successful results. Before you begin, a maintenance plan should be in place to ensure that this crucial portion of the project is not later neglected.

For one to two years after planting, it will be necessary to carry out an aggressive weed control program to ensure the successful establishment of the meadow. As the process of ecological succession would suggest, the first year meadow often contains significant numbers of annual weeds while the perennial wildflowers and grasses are slowly developing underneath. This is to be expected and, if managed properly, is not a problem. Mow the meadow every six weeks to a height of 4–6 inches and you will not only prevent the annual weeds from seeding, but also ensure that the young perennial plants growing below the mow height receive enough light for strong establishment. These perennials will emerge the following year far stronger than if they had been buried under 4 feet of annual foliage the first year. This is why the inclusion of annual wildflowers in the seed mix can be detrimental to the long-term health of the planting. Annual wildflowers are included for their ability to bloom the first year, and in order for this to occur you will be prohibited from mowing. This will allow annual weeds to go unchecked and deprive the emerging perennials of the light needed for optimal growth.

Although most common garden and lawn weeds like common purslane (*Portulaca oleracea*) and dandelion (*Taraxacum officinale*) require no action—the meadow plants will outcompete and suppress them on their own—it is important to take action against those weeds that can pose a threat. Herbaceous weeds like Canada thistle, mugwort, or reed canarygrass (*Phalaris arundinacea*) can be controlled with spot herbicide applications directly to the foliage of individual plants. Woody plants can be treated the same way; or you can cut the stems and paint the cut stubs with herbicide, as Frank Egler did with unwanted trees at Aton Forest. Given the localized nature of either of these application methods, and the minute amounts of herbicides that are applied, I don't believe that any significant environmental effects result. Manual weeding, repeated cutting at the base, or an additional mowing immediately following the most active growth period of the problem weed are methods that can be used if herbicides are not acceptable. Ideally, one weed control walk-through per season (winter excluded) should keep the meadow on a good trajectory.

Tall drifts in a designed meadow frame the entrance to an orchard, accentuating the passage from enclosure to openness.

During the second year, biennials, such as blackeyed Susan and spotted beebalm (*Monarda punctata*), will dominate the meadow and provide the first flowers. A decrease in weeds may be noticed, but you should still consider this the development phase, and seasonal monitoring and spot control should be continued. A mowing in late spring to control cool-season grasses and weeds may be needed, but by and large the meadow now and going forward will be cut once per year in late winter.

Once you have stopped cutting during the growing season, you can begin to mow paths, sitting areas, and lookout points within the meadow. Most often I simply mow the meadow vegetation to make these paths, but you can create a more refined look by sowing them with a low, clumping grass, such as sheep fescue or hard fescue. Avoid sowing mat-forming turf grasses like Kentucky bluegrass as their roots will creep into the meadow and create an unsightly edge.

By the third year the native meadow plants should be fairly dominant on the site and able to resist weed invasion with much less management. The first late-stage species like eastern purple coneflower and wild bergamot (*Monarda fistulosa*) will produce their first blooms. Maintenance will consist of one spot control visit per season, mowing of paths as needed for comfortable walking, and one global mowing at the end of winter. In place of the winter mow, you can also perform a controlled burn. Most experts feel that this method is superior, as many prairie species historically evolved with burning from accidental lightning fires and intentional fires set by Native Americans prior to European settlement. In our current cultural and regulatory climate, this can be difficult to pull off. But like allowing the return of predators to control exploding deer populations, which is now under consideration in some areas, this could change.

There is another optional procedure that may be employed for aesthetic reasons. Often the plants on the very edge of the meadow grow noticeably taller than those in the interior. This is due to the additional light they receive on the side where the lawn, driveway, or other short surface meets the taller meadow. The taller plants can block the view into the thriving meadow that you worked so hard to achieve and be very annoying. If so, cut the edge vegetation down by about two-thirds. It will resprout and flower but never attain the same obstructive height as it was previously.

The first several feet along the edge of this meadow were cut back to prevent the edge vegetation from becoming too tall and blocking the view. This moderate disturbance took succession backward, and short-term species like blackeyed Susan re-emerged. Once I realized the potential for this type of disturbance, I began to use it on other projects to achieve particular effects.

STEPS TO MAKE A MEADOW

Preparation of Areas to Be Converted to Meadow

The procedures outlined here are appropriate for a typical area to be converted to meadow. However, sites dominated by many invasive weeds may require additional preparation over a longer period of time.

NINE WEEKS PRIOR TO SEEDING: delineation
Delineate the meadow boundaries using flags, stakes, or marking tape.

EIGHT WEEKS PRIOR TO SEEDING: first herbicide application
Apply appropriate herbicide as directed on the label and at appropriate concentrations to kill existing turf grass (and weeds, if any).

FIVE WEEKS PRIOR TO SEEDING: mowing
Scalp-mow dead growth to 1 inch or shorter to allow for germination of the remaining weed seeds. Rake off excess plant refuse so that there is approximately 50 percent soil exposed. Leave dead roots and crowns to stabilize soil prior to seed germination.

TWO WEEKS PRIOR TO SEEDING: second herbicide application
Apply appropriate herbicide as directed on the label and at appropriate concentrations to kill remaining turf grass and any newly emergent weeds.

TWO TO THREE DAYS PRIOR TO SEEDING: mow dead material
Inspect the site to ensure that no further herbicide treatments are necessary. If growth is completely dead, then mow the area as low as possible (approximately 1–2 inches). Rake off excess plant refuse, but be sure to leave dead roots and crowns to stabilize soil prior to seed germination. Mow and rake as needed until 50 percent of the soil is exposed.

Seeding Procedures

Spring, early summer, and autumn are the appropriate seasons for seeding throughout most of North America, with the exception of the subtropical extreme south and the arid Southwest, where seeding must coincide with periods of increased soil moisture and relatively mild weather.

Spring through early summer
In areas of heavy weed infestation, seeding in late spring to early summer can be a good option, as most weeds are past their active growing season. Many of the meadow flowers and grasses are heat adapted, germinating and growing well at this time of year. In the event of a summer drought, the seed will remain dormant but viable until rainfall resumes.

Autumn
This is effectively a dormant seeding, so the plants will not germinate until the following spring after sowing. This should not be performed on sloped areas where seed can be washed away during autumn and winter rains and times of snowmelt.

Meadow encompasses this lakeside property in Connecticut. A meandering path leads from the house to the lake.

METHOD 1
No-Till Broadcast

This method should be used in areas where it is not possible to bring in a drill seeder, such as wet or very steep areas, or in small sites where bringing in large equipment is not appropriate.

1. Scarify soil to depth of approximately ½ inch using appropriate equipment or by hand. Do not till because this will increase weed presence due to the release of dormant weed seeds in the soil.

2. Segregate the seeds according to mix and seed size. Fluffy, small, and grain-sized seed may be combined for hand broadcast and will henceforth be referred to as "large" seed. Very fine seed must always remain separate.

3. Inoculate leguminous seed. Place the legume seed in a container and add a small amount of sand, sawdust, or other fine-textured bulking material. Moisten the mix until slightly damp. Add the inoculant and mix thoroughly. Mix this inoculated seed into the large or very fine seed as specified.

4. Bulk up the large seed by mixing it with an inert bulking material such as sawdust, sand, kitty litter, or vermiculite. Use enough bulking material to allow seed to be evenly distributed over the entire area. Slightly dampening the bulking material will help the seed distribute more evenly in the bulked-up mix.

5. Hand broadcast the large seed mix evenly over the surface of the designated area. Use a pattern and method of dispersal that reduces the chances of using up the mix prematurely. Subdividing and marking the area into sections can facilitate the process.

6. Rake seed lightly into the soil to a depth of approximately ¼ inch using appropriate equipment.

7. Bulk up the very fine seed in the manner described above and broadcast this mix over the surface. Do *not* incorporate into the soil, as these species need light to germinate and will not do so if buried in soil.

8. Tamp seed for proper seed-to-soil contact with a roller, drag, or other appropriate tool.

Mulching the planting is usually not required, as the remaining dead stubble and plant refuse will perform this function. If there is not sufficient mulch from these sources, spread a thin layer of weed-free straw, such as salt-marsh hay or winter wheat straw. Mulch should leave approximately 50 percent of the soil exposed to sunlight.

Watering is usually not required. If no rain falls for an extended period after planting and a source of water is readily available, application of up to 1 inch of water per week (including any rainfall) for a period of six weeks will greatly facilitate meadow establishment.

METHOD 2
No-Till Drill Seeder

A no-till drill seeder is equipment adapted specif-ically for native grasses and flowers that typical agricultural drills cannot adequately accommodate, due to the variety of seed sizes and textures of native plant species.

1. Scalp-mow all remaining dead growth to expose the soil to light.

2. Segregate species in each seed mix and handle as follows:

   Fluffy Place in fluffy seed box

   Small Place in alternate sections of the small flower box

   Very fine (too small to pick up individually) Bulk up with sand or vermiculite (not sawdust, as it may jam the seeder) to the same volume as the small seed, and place in remaining sections of the small flower box

   Grain Place in grain box

3. Remove feeder tubes from the small flower box sections containing very fine seed. This allows the very fine seed to fall directly onto the soil surface without being cut in by the machine, thus receiving the light necessary to germinate. Do not remove the feeder tubes for sections containing small seed.

4. Calibrate the seeder. This requires some trial and error, as every seeder is slightly different. Drive some test runs to see how quickly the machine deposits seed and adjust accordingly. Never put all the seed into the machine at once. Always divide the area and the amount seed accordingly, which eliminates the possibility of accidentally running out of seed partway through.

5. Ensure that the drill seeder is not depositing seed too deeply into the soil. The disks should be cutting into the soil approximately ⅛ to ¼ inch deep. During the trial run to calibrate the seeder, examine the slits made in the soil to ensure proper depth. Some seed should be evident on the surface of the soil and some will be tucked into the slits.

6. Tamp seed for proper seed-to-soil contact with roller, drag, or other appropriate tool.

Mulching the seeded area is usually not required, as the remaining dead stubble and plant refuse will perform this function. If there is not sufficient mulch from these sources, spread a thin layer of weed-free straw, such as salt-marsh hay or winter wheat straw. Mulch should leave approximately 50 percent of the soil exposed to sunlight.

Watering is not usually required. If no rain falls for an extended period after seeding and if a source of water is readily available, application of up to 1 inch per week (including any rainfall received) for a period of six weeks will greatly facilitate meadow establishment.

GROWING PAINS

Early in my career I visited Bob Swain, a landscape contractor with whom I often worked. His company primarily took on large-scale native restoration projects, and his office was located in the outer coastal plain of central New Jersey, an ecoregion where large areas of pine barrens vegetation predominate. I walked from the parking area to the front door of his office on a boardwalk that cut through a thicket of inkberry holly (*Ilex glabra*), a ubiquitous member of the pine barrens plant community.

Bob had planted the thicket, but these inkberry hollies were unlike the ones I had seen in gardens elsewhere, which were clump-forming, upright shrubs. Bob's hollies formed a solid network of vegetation, a wavy series of connected mounds that led all of the way to his office door. I was struck by the stout yet airy expanse and the fact that it resembled one big, connected plant. I had used inkberry holly before, but it never looked like this.

I asked Bob where he got the plants. He told me they came from the nursery next door. I asked if he had pruned them in any particular way. He hadn't. I asked if they were some kind of broad, spreading selection or cultivar. They weren't. "Okay." And we went on to another topic.

Fast forward about twenty-five years, and I am at a conference in Ithaca, New York, speaking with the very knowledgeable native plant grower Dan Segal. During our conversation, he noted that inkberry holly is naturally a clonal shrub with a creeping pattern of growth, but that it seems to lose this habit when propagated asexually by cuttings. This was an important

bit of information because virtually all inkberry hollies sold by landscape nurseries are cultivars, plants selected for compact growth, glossy foliage, or other ornamental qualities, and these cultivars must be propagated by cuttings to preserve the characteristics for which they were selected.

He said the same appeared to be true of other clonal spreading native shrubs, such as summersweet (*Clethra alnifolia*), winterberry holly (*Ilex verticillata*), and chokeberry (*Aronia* species). They are all clonal spreaders in the wild and clump-formers in the garden. Like the inkberry, they are almost all sold exclusively as cultivars by ornamental nurseries.

Immediately a switch went off in my head, and I thought of Bob Swain's boardwalk landscape in the pine barrens. I now knew that a nursery next door to Bob's was owned by Don Knezick, a pioneer grower of seed-grown, straight-species native plants. So the plants that Bob got from "the nursery next door" were seed-grown plants, not cultivars that had lost their clonal ability. That's why Bob's inkberry hollies formed such a beautiful thicket: they had come from Don Knezick's nursery.

Why didn't I realize that back then, when I had first seen that planting? It should have been obvious. Those plants were behaving differently because they *were* different. Although I already considered myself an

Meadow was originally to be planted through this birch grove. After observing the presence of native shrubs, including two native spireas, and gray dogwood, I decided to manage for shrubland instead, removing invasive plants and allowing the native shrubs to take over. No planting needed.

This cultivar of inkberry holly, propagated asexually by cuttings, has lost its ability to spread by roots, becoming unnatural in appearance.

advanced practitioner of "native design" at that time, I still had one foot and four toes in the world of horticulture instead of one foot in each world—ecology and horticulture—as I hope is the case now. In thinking about those inkberries, I hadn't considered the functional role native plants play in the landscape or the strategies they employ to fulfill those functions. Precisely because of their thicket-forming characteristic, clonal shrubs are some of the most weed suppressive, soil-stabilizing, and wildlife-friendly plants you can put in your garden—unless you plant a cultivar.

I'm not against cultivars. A cultivar of inkberry holly is a good choice for a small garden where a clonal spreader would swamp its neighbors, but it isn't a good choice to colonize a large sweep of lawn being converted to a rain garden.

A common complaint about inkberry holly cultivars is their tendency to "limb up" and reveal exposed stems over time. Thinking about it now, I realize why. This species evolved to expand, with new perimeter shoots continually covering the edges of the plant as the interior ones stretch. A graceful appearance like the one I observed on Bob Swain's boardwalk is the result. But with a cultivar, expansion is truncated, and no new outer shoots emerge to cover the stretching inner stems. The growth pattern is interrupted, and so the plant looks leggy. I haven't uncovered any scientific studies to confirm this particular relationship between propagation method and growth patterns. I have observed this phenomenon many times, however, in nature and in gardens, and it now influences plant sourcing for my firm's projects.

Fortunately, horticulture has learned to embrace native plants since Bob Swain planted Don Knezick's seed-grown inkberry hollies, and the following story about Don's nursery vividly illustrates that. I first met

Don staffing a booth at a landscape trade show. He had just started his nursery and was trying to drum up some business. But this was a show for high-level professional designers, and these prospective customers uniformly peered at his display of spindly little 1-gallon shrubs as if thinking, "Why would anyone want to plant those?"

They didn't understand that to make large area plantings affordable you need smaller plants. They didn't understand that full bushy plants are spoon-fed in the nursery and will be in for a huge shock when they hit the real world, especially the rough and tumble of a wild landscape. They certainly didn't realize, nor did I, that whether a plant was propagated from seed or by cutting had any significance to its future growth.

At the trade show Don looked like the loneliest man in the world, and I doubt he did much business there. Today, however, he runs a huge operation. He sells plants throughout the Mid-Atlantic states and is recognized as a leader in the field, and not just by restoration ecologists.

Native plants are no rarity now. But visit a garden center or a traditional wholesale nursery and try to find a straight-species, seed-grown inkberry holly. You can't. If you want a native clonal shrub to colonize a large bank aggressively, ironically you will be limited to those species like sweet fern, gray dogwood, and winged sumac (*Rhus copallinum*) for which no "improved" cultivar is yet commonly available to supplant the wild-type plants.

Improved selections of these shrubs are surely on their way, however, and when they arrive, the improvements will likely take the form of a visual characteristic such as shinier leaves, dwarf habit, or altered flower color. Overall, I think that in the plant selection process horticulture has placed too much emphasis on the ornamental and not enough on the functional. I don't need a purple-flowering sumac. I need one that can cover a bank quickly, hold back the weeds, and provide a sheltering thicket for birds. So when sumac 'Plum Fizzle' hits the trade, I may plant a few for interest. But when it comes to my steep, problematic bank, it won't make the cut. What about the aesthetics? As far as I'm concerned, an autumn view of good old, seed-grown blazing red sumac in a sea of golden Indiangrass is about as good as it gets.

top Periodically moving this path away from the gray dogwood thicket allows it to expand into the meadow, creating a visual and ecological transition between the tree grove and the meadow. Once the shrubs have extended to the desired point, the path alignment remains fixed, holding the shrub thicket from expanding further.

bottom Not only does this path function as management tool to control the shrubland edge, but it is placed to highlight the stately tree in the distance.

CREATING SHRUBLANDS

Planted meadows are increasingly common in the landscape, as are woodland gardens and reforestation plantings. But the intermediate successional stage, the shrubland, is most notable by its absence. By shrubland, I am not referring to the woodland shrub layer. Rather, I mean an open, sunny area in which shrubs predominate.

Such vegetation is valuable for several reasons. As Frank Egler demonstrated in Aton Forest, it is perhaps the most stable vegetation type. Once well established, a shrubland can persist with minimal management for decades. Indeed, in the form of chaparral, shrubs dominate large areas of southern California, just as sagebrush (*Artemisia* species) covers wide expanses of western rangelands.

Although shrubland is the intermediate stage between meadow and woodland, there is no distinct line, no characteristic moment, when meadow transforms to shrubland and when shrubland transforms to forest. Given that shrubland is commonly the middle stage, it often intermingles with its predecessor, the meadow, and its successor, the forest. This is actually one of its strengths from the aesthetic perspective. This intermingling gives shrubland a dynamic visual character. Looking at a shrubland feels like looking at change. This mixed character can also make it more complicated to manage, as it is neither fish nor fowl. You can't mow the whole thing like a meadow or manipulate light as in a woodland. Consequently, shrubland planting and management have their own set of techniques, which we will explore later in this chapter.

Shrubland also completes the portfolio of back-yard wildlife habitats. Just as some birds and other creatures prefer woodland and some grassland, many require the low, dense cover of the shrubland. Often, in fact, wildlife considered woodland or grassland dwellers also depend on the ecotone edge—the place where the habitat changes—and that tends to be dominated by shrubs.

If nothing else, shrubland is an aesthetic opportunity. A lot of really beautiful shrubs are native to North America. Planting them in large masses, as in a naturally occurring shrubland, only increases their visual impact.

As with a meadow, it's not just the plants but the patterns. Transitioning from woods to meadow with periodic peninsulas of native shrubs can transform a rectilinear remnant farm field into a dynamic composition that evokes both its pastoral history and its wild future. Placing isolated drifts of the same shrub species out in the meadow can take the meadow-to-shrubland process a bit farther down the timeline, and further blur the lines between the landscape's basic building blocks of woodland, shrubland, and meadow.

Given that the foliage of most meadow species is below eye level and that the branches of most trees are above, shrub masses are the most useful for defining space. Screening unsightly views, framing desirable ones, and creating garden rooms—or in our case, natural area rooms—are all best accomplished with shrubs. Fostering a feeling of surprise is a common technique for enhancing the drama of a garden, and what could be more surprising than following a path toward what appears to be a solid shrub mass in the distance, only to find out when you get there that it is hollowed out and you can actually pass into

As a driveway passes through an established woodland, aggregates of the rhizomatous shrub bottlebrush buckeye were added for visual interest. A gap was left to highlight a mature clump of trees.

it? Once inside, maybe you encounter a hammock. Climbing in, you look up and notice that that the top of the shrub drift forms a perfect frame for the sky, a natural masterpiece that will be different every time you look at it.

I spoke earlier of creating peninsulas of shrubs that extend from the woods' edge into the meadow. Run a path from the woods' edge through the center of that dense shrub thicket, and you will create an entry to the meadow that opens suddenly and dramatically.

Too much shrub thicket, however, can also be intimidating. I once asked a landscape architect friend, who was also a talented painter, to create a painting in which the vantage point was from the interior of a shrub thicket. Evidently, he did not find the thicket a comfortable place to be, as he painted it with an opening that provided a view of a lake. Placing paths through alternating areas of thicket and meadow can alleviate this claustrophobic aspect by creating a rhythmic interplay of closed and open space, a common garden design technique.

CATEGORIES OF SHRUBS

In terms of their growth habits, there are two basic categories of shrubs: rhizomatous and stoloniferous shrubs, which proliferate through vegetative means, and clump-forming shrubs, which reproduce via seed.

Rhizomatous shrubs are those with dense, expanding root systems. Typically new shoots or suckers emerge from the ground as the roots extend outward, gradually turning the individual plant into a broader clump or thicket. In this way, rhizomatous shrubs often (although not always) form monocultures due to the competitive nature of their dense clonal root systems. Examples of such rhizomatous shrubs are gray dogwood, bottlebrush buckeye, and sumacs.

If the conditions are favorable, low-growing species of this type of shrub can form wonderful groundcover compositions. These species are often associated with dry habitats. Examples of this sort of rhizomatous shrubs are sweet fern, spreading junipers, and lowbush blueberry, which provided the materials for Frank Egler's shrubland at Woodchuck Hill.

Many contemporary gardeners complain that lowbush blueberry is painfully slow to establish when planted. Historically, fields of this plant were burned periodically, particularly in New England, which increased vigor and fruit production. The early European colonists probably learned this from Native Americans, who burned extensively to manipulate vegetation. In contemporary times, local codes make burning a whole field problematic, but burning individual clumps of bushes or a small patch with a blowtorch may be allowed and produces the same effect.

Rhizomatous shrubs are valuable for stabilizing erosion-prone soils and for bank plantings, particularly in spots where it is too steep to mow, even once a year, and hence meadow is impractical. They are also the best of shrubs for weed suppression, typically proving more competitive even than meadow. This fact was well illustrated by Frank Egler and Bill Niering's work on power line rights-of-way. Hired by power companies to research vegetation types that would suppress tree growth under power lines where mowing was impractical, they found through experimentation that dense shrub cover was most effective, requiring only periodic tree sapling removal as management, and that with decreasing frequency.

In general, sumacs are some of the most effective shrubs for this purpose. One notable exception, however, is staghorn sumac, which has elongated stems that allow more

Box huckleberry, a clonal shrub, can form large colonies if provided with the right conditions. This clonal trait can be exploited for aesthetic and functional purposes in designed landscapes.

AN ANCIENT GIANT

Linked together by creeping roots, a whole thicket can consist of a single plant. According to a report by Rob Nicholson in the newsletter of Harvard University's Arnold Arboretum, an extended colony of box huckleberry (*Gaylussacia brachycera*), an evergreen subshrub, in Perry County, Pennsylvania, that once covered more than 100 acres is composed of clones of a single individual. Genetic analysis has shown the colony to be between 1000 and 2000 years old, making it possibly the oldest known living woody plant east of the Rocky Mountains.

light to the ground underneath. As a result, although rhizomatous, it is less effective at suppressing tree growth under its spread. Of course, this makes staghorn sumac a good choice if you do want the shrubland to be replaced by forest over time. In general, though, rhizomatous shrubs are an effective type of vegetation for preventing forest regrowth.

Low-growing shrubs can provide a mowing-free cover for dry banks and dry, poor soils in general, and other types of clonal shrubs can fulfill a similar function in wetlands, where mowing is also difficult or impractical. Notable examples include silky dogwood (*Cornus amomum*), common buttonbush (*Cephalanthus occidentalis*), summersweet, winterberry holly, and red chokeberry (*Aronia arbutifolia*), as well as steeplebush spirea and white meadowsweet. These shrubs are relatively common garden shrubs, with the exception of the native spireas. The latter should have their place, too. Both spirea species are low-growing, bulletproof plants with spike-shaped pink or white flowers. Their brownish black stems are particularly stunning against the winter snow.

Stoloniferous shrubs, many of which are groundcovers, include bearberry (*Arctostaphylos uva-ursi*) and eastern teaberry (*Gaultheria procumbens*). These species can weave in and among other vegetation, rooting wherever they find a bare spot. Some shrubs also have the ability to root at points where a branch meets the ground and can be used similarly to rhizomatous shrubs.

Masses of fragrant sumac, *Rhus aromatica* 'Gro-Low', a dense, low-growing, rambling shrub, intermingle with meadow vegetation along a driveway entrance. The dense matrix suppresses weeds while helping to stabilize the embankment.

left This shrubland in Maine is a patchy matrix of common juniper and northern bayberry (*Myrica pensylvanica*) underlain with grasses and lowbush blueberry. Eventually, if not managed as shrubland, the area will become spruce-birch forest like that visible in the background.

right The open branching and lack of lower foliar growth on staghorn sumac allows trees to penetrate, making it a good shrub to foster forest and a poor one to create a stable shrubland.

Silky dogwood, a stoloniferous shrub native to eastern North America, thrives in moist to wet natural areas, providing excellent erosion control. The creamy white flowers in late spring to early summer give way to berry-like drupes highly sought after by birds.

Clump-forming shrubs, the second broad class of shrubs, are those that form relatively compact, individual plants and that reproduce by seed, rather than by suckering. Examples include viburnum, highbush blueberry, and sweetshrub (*Calycanthus* species). Their compact, nonspreading pattern of growth makes clump-forming shrubs more suitable for intermingling with other shrubs or with perennials and grasses.

These shrubs are sometimes criticized as being leggy, but this is a characteristic that makes them more suitable for intermingling with herbaceous plants. If you do intermingle in this fashion, it means that there will be two layers of vegetation through which weeds have to penetrate to colonize the area. Besides, I find that the leggy stems of some shrubs such as highbush blueberry can be quite architectural in character and add a dramatic visual element to the landscape.

SHRUBLAND PATTERNS

Shrublands fall into three general patterns: solid shrubland, mixed oldfield, and mixed woodland edge. The type of shrubland that expresses itself on a particular site depends partly on underlying conditions but also its history, both natural and of human usage.

In solid shrubland, the coverage is nearly 100 percent shrubs, often with a high proportion of rhizomatous species. At times, such a shrubland can be virtually a monoculture because some rhizomatous shrubs, including gray dogwood, are such aggressive spreaders that they squeeze out everything else.

In a mixed oldfield, shrubs intermingle with herbaceous plants and possibly with pioneer tree species. If trees are present and allowed to grow unchecked, they will eventually shade shrubs and weaken them, setting the stage for the encroachment of more trees and eventually leading to succession to forest.

Berry-forming shrubs that are distributed by birds, such as blueberry, viburnum, and winterberry holly, appear seemingly randomly throughout the area of mixed oldfield. In fact, although it's difficult to predict exactly where a bird will defecate, you are likely to find heavier concentrations of these plants below convenient perches where birds spend time, such as trees and telephone poles. If you want to recruit berried shrubs in a particular spot, you can plant a tree there. Alternatively, you can erect a pole for a perch or even assemble deadwood into a temporary sculpture à la Andy Goldsworthy, the British artist who uses natural objects to construct his wonderful creations. Such a piece can furnish ephemeral artistic expression while also serving an ecological function until it decays. Perches are less suitable for meadow plantings, where the shrubs they encourage can become management problems.

The composition of mixed woodland edge, a combination of shrubs, herbaceous vegetation, and trees, is the same as mixed oldfield, but the shape is more linear or may be a series of peninsulas extending out from the woods into the meadow or lawn. The peninsulas enlarge naturally as rhizomatous shrubs at the edge of the woods expand out into the adjacent open landscape or by the recruitment of clump-forming shrubs by seeds that fall from their branches.

Staghorn sumac colonized this roadside embankment, functionally stabilizing the steep bank while providing a flaming autumn display.

Shrub thickets can be formal in arrangement and need not be limited to natural areas. The plants don't care, nor would most wildlife. Here, highbush blueberry forms an allée, a traditional landscape motif, in keeping with the property's Edwardian house. Depending on the vantage point, this landscape can appear ultra-formal or wild.

left Gray dogwood will likely overcome the goldenrod in this composition because shrubs generally follow herbs as ecological succession advances.

MANAGING A SHRUBLAND

You can direct the successional trajectory of a field toward shrubland by applying or withholding various management techniques.

Cutting the field will always favor the herbaceous species. Perennial herbs and grasses are accustomed to regenerating growth after being cut. It's in their DNA, and they do it every winter. Shrubs, on the other hand, do not lose and replace top growth as part of their normal life cycle. They invest a lot of energy in creating their woody, more permanent stems and twigs, and to rob them of this with a mower sets them back seriously.

Accordingly, withholding mowing favors shrubs and eventually trees. But what of the weedy plants, woody and herbaceous, that are so often present in open field landscapes? Your choices for controlling them will generally be either periodic selective cutting or selective herbicide application. In both cases, the uncut or unsprayed native shrubs, either existing or planted, will be favored and may be expected to dominate over time. Even if herbicide is being used, it is often best to cut the unwanted plants first, then either paint the stems or spray the low re-emerging growth. Cutting essentially creates a height differential that allows you to apply the material to the short weeds without affecting the uncut, taller native shrubs. This is particularly effective when the desirable and undesirable plants start out intermingled.

MANAGING AN OLDFIELD

An oldfield contains both meadow and shrubland, and so mowing must be both applied and withheld. Where you mow or don't mow will determine the ultimate pattern that the shrubs and herbs will express.

Begin by determining the desired balance of areas you want covered with shrubs versus areas covered with herbaceous plants. You may want to preserve the existing distribution, move the field back in time toward a higher percentage of meadow, or allow the process of succession to proceed and increase the shrub component. Next, delineate where each shrub mass will occur. You can use ribbon, stakes, or flags to accomplish this. At this point you are designing in the field, and the following are some approaches to guide you in this process. Usually you'll find yourself combining all of these approaches.

Go with the flow. Place your drifts where shrubs already have a stronger presence than the herbaceous plants. This is the path of least resistance and is likely to result in achieving the desired composition relatively quickly.

Concentrate on the edges. Concentrating shrub masses around the perimeter of a space is generally appropriate where meadow meets woodland, and will keep a large central area open for meadow.

Let your sense of aesthetics guide you. This is where art and natural processes intersect. You can arrange the shrub patches as you would in a thoughtfully designed garden, except here you are deciding where to remove shrubs, not plant them.

This path both highlights a specimen tree and marks the edge between a designed meadow and shrubland. Regular mowing prevents the rhizomatous native shrubs from spreading into and colonizing the meadow.

Having established your design, mow accordingly. The shrubs that aren't located in the islands will be discouraged by the mowing and soon fade away. These areas can then be treated as described in the preceding meadow chapter. The remaining shrub islands, which may have weeds, desirable shrubs, or both, can be treated as described in the previous "Managing a Shrubland" section.

You will likely be amazed at the striking visual effect that this wild, yet highly defined landscape produces. Creating distinct areas of shrub and meadow has an important practical benefit as well. In a naturally occurring oldfield, some of the shrubs may form isolated patches, particularly those that expand clonally by rhizomatous roots. Others, however, may be scattered individually throughout the field. Mowing around sizable patches of shrubs is easy and certainly much more fun than mowing turf back and forth. But even that tedious job is better than dizzily circling around hundreds of individual shrubs. Creating distinct areas of meadow and shrubs may not perfectly reflect the natural patterns of an open field landscape, but such targeted mowing may be the best way to arrest succession and preserve this particular moment in the field's successional process.

But what if you don't want simply to maintain the status quo? Maybe the winged sumac that you are managing for occupies 15 percent of the field, and you want it to own 25 percent. If you wish to encourage the spread of this rhizomatous shrub, you will need to adjust your mowing. Mow only to within 24 inches of the shrub's edge to allow new shrub shoots to emerge around its periphery. Then, when the new shoots have emerged, follow up by mowing the 24-inch unmowed zone around the shrub, but this time with the blade raised high enough to clear the tops of the newly emerging shoots. This helps the new shoots compete against the vegetation into which they are expanding. It will also give this battle zone between meadow and shrubs a neater look while it's still in the process of sorting itself out. As the shrubs expand outward, repeat this pattern of mowing until the shrubs have reached the maximum area that you want them to cover. Then mow right up to the edge of the shrubs to stop the expansion.

Sometimes the once-a-year mowing you apply to the surrounding meadow areas in the oldfield will not be enough to contain a particularly aggressive shrub patch. This was the case with gray dogwood on the property in Dutchess County, New York. In that case we mowed a circular path around the perimeter of each dogwood island and connected them to the mowed path system that wanders through the meadow portions of the field. Now, whenever the paths are mowed, the dogwood's spread is checked. Doing this also has opened access to more areas in the field.

Finally—bear with me here, this technique is optional and applied only once every eight or ten years—there is rejuvenation cutting of the rhizomatous shrub islands. One of the most attractive aspects of rhizomatous shrubs is the mounded form they naturally adopt. Almost formal, these dense mounds can dramatically punctuate the field with a striking contrast to the delicate grasses of the surrounding meadow. If you mow repeatedly right to their edge, you will gradually destroy this look as the peripheral shoots grow taller and no new shoots emerge around the edge. To regain the mounded form without allowing the shrub to expand, cut it back to the ground. It will quickly re-emerge and regain its former mounded architectural character.

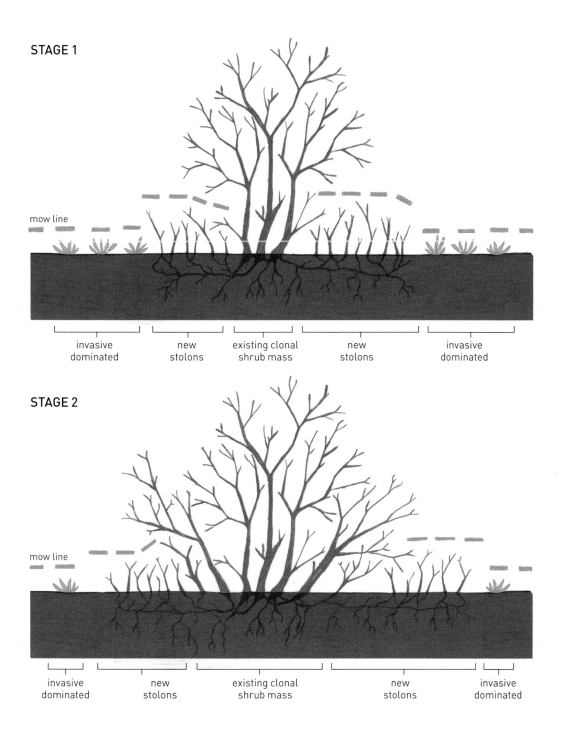

STAGE 1

mow line

| invasive dominated | new stolons | existing clonal shrub mass | new stolons | invasive dominated |

STAGE 2

mow line

| invasive dominated | new stolons | existing clonal shrub mass | new stolons | invasive dominated |

This diagram illustrates a process for fostering the expansion of clonally expanding shrubs. A selective height mow regime is used to allow native shrubs to expand and displace adjacent weed species over time.

MANAGING TREES IN THE SHRUBLAND

Whether you have elected to allow the entire field to succeed to shrubland or just selected patches, you will have another decision to make. If you want these shrubs to succeed to forest, do nothing and natural processes will accomplish your goal. If you want shrubland to be the long-term composition, however, then removal of individual trees should be on your to-do list.

The most effective way to do this is to cut the trees at the base and then paint the raw top of the stump with an appropriate herbicide. Because this sort of herbicide application is so closely targeted, it involves minimal use of chemicals. This was the technique that Frank Egler used so successfully at Woodchuck Hill and in his work under power lines. Whether you even need this step is largely determined by the density of the shrubs and the resprouting ability of the cut trees; for example, maples (*Acer* species) often resprout, whereas eastern redcedar and eastern white pine (*Pinus strobus*) do not. This, incidentally, is a self-reinforcing kind of management: the longer that you persist in removing trees, the thicker the shrub cover will become and the fewer trees will be able to struggle up through it.

PLANTING A SHRUBLAND

When seeding a meadow, you must eliminate all the prior existing vegetation first or it will suppress the germination of the meadow seeds. Shrublands, however, are generally planted with live plants, so that you can often insert them right into the existing vegetation. Shrubland is a later successional stage that is better able to cope with herbaceous surface vegetation, which means you have a choice. You can, if you wish, follow the more typical garden approach and clear the area, plant the shrubs, and then mulch around them. Alternatively, you can insert the shrubs into the existing vegetation and then manage the area in such a manner as to favor the shrubs.

Another important aspect of planting a shrubland is the way in which you combine shrubs. If you mix the two types of shrubs, rhizomatous/stoloniferous and clump-formers, in the same mass, the clonal species will typically overwhelm the clump-formers. Consequently the two types of shrubs should be segregated.

An exception to this rule is if you plant tall clump-formers with low-growing rhizomatous shrubs: southern arrowwood and lowbush blueberry, for example. In such a case, the two types are occupying different spatial niches within the landscape and should collaborate rather than compete.

Generally, it is fine to mix different species of clump-formers within the same mass. This will increase diversity of the landscape, benefiting wildlife and making the shrubland richer visually. In nature, rhizomatous shrubs often grow in single-species masses.

Planting them in this pattern is generally advantageous, although you can mix rhizomatous species if they are of equal competitive levels.

The natural distribution pattern of both clonal shrubs and those with gravity-dispersed seeds is to form large isolated patches, with the growth moving out slowly from the mother plant. Arranging these sorts of shrubs in this fashion in the landscape will not only create an authentic appearance, but will make meadow areas between the shrubs well defined and easy to negotiate with a mower. This is a perfect example of the concept of designing for management. You may treat the resulting shrub patches as the permanent composition, or you may allow them to expand into larger patches or even to colonize the entire field. Because spontaneous expansion of the shrub patches can be slow, the larger the planted patches and the closer you space the shrubs, the sooner you'll achieve any desired spread.

Shrubs distributed by birds tend to produce individual specimens scattered throughout a field. This scattershot pattern helps to accelerate shrub expansion but makes managing the expansion more difficult, as mowing between the shrub patches can be more complex.

In this shrubland, where the intent is to keep it at this successional stage, tree saplings can simply be cut at the base and the cut stump painted with the appropriate herbicide. Alternatively, if herbicide is not used, any resprouting growth able to re-emerge among the dense shrubs can be cut again until the tree is too weakened to compete with the surrounding shrubs and drops out.

SEDGES IN THE LANDSCAPE

North America is home to a tremendous number of
sedges, with species living in almost every type of habitat. They are keystone species in many wetland communities, playing a critical role in maintaining the ecological structure there. Although these particular species,
such as yellowfruit sedge (*Carex annectens*), rarely
appear in residential gardens, they are often planted
from seed in large-scale wetland restoration projects.

Many sedge species colonize woodlands as well,
and some form beautiful carpets in gaps and areas with
filtered light. Pennsylvania sedge and Appalachian
sedge (*Carex appalachica*) are two examples. Though
delicate in appearance, they form remarkably tough
woodland groundcovers. Because they use the limited
light resources available in the forest to expand clonally
by rhizomatous roots, they produce little viable seed.
Consequently, they can only be planted as live plants in
the landscape, which makes them cost prohibitive for
large-scale landscapes.

Recently, however, our firm began searching for
lesser-known woodland sedges that *could* be sown successfully from seed. We didn't find the information we
needed in the horticultural literature, and so we turned
to restoration ecology and botany publications. This
uncovered some promising species, including slender
wood sedge (*Carex riparia*), plains oval sedge (*Carex
brevior*), limestone meadow sedge (*Carex granularis*),
troublesome sedge (*Carex molesta*), and eastern star
sedge (*Carex radiata*).

With these species we formulated a seed mix for
a property that included a woodland area of approximately 2 acres. Two things prevented us from seeding
the entire space. First, it would be too big a commitment
for an experiment, particularly using species with which
we had little experience. Second, after pricing the seed
mix we found it would cost almost as much as live
plants. Instead we decided to go a more affordable route
and plant very small patches along the woodland paths
and see what transpired.

The patches were planted in autumn of 2013. By
the following autumn extremely dense clusters of tiny
sedge seedlings had formed in each patch. Obviously, I
was encouraged by the presence of sedge plants, but I
was equally encouraged that the density of those plants
was far greater than was needed for colonization. This
meant that I could cut back significantly on the seeding
rate and that a mass seeding of the entire woodland was
likely to be not only successful but affordable.

Our client was involved in this entire process, and
in autumn of 2014 he gave us approval to seed the entire
2-acre woodland. As I am writing this in early spring of
2015, I can't report on the outcome of phase two. I am
confident that it will be successful, however, because
the planting was preceded by research of the ecological
literature, discussions with our seed suppliers, and, most
importantly, experimentation in the field.

Small-scale field experimentation prior to embarking
on major efforts over large areas is something that I feel
should be more common. This is particularly true when
lawn replacement and natural area restoration causes us
to transform areas at a large scale.

I can think of one instance where preliminary
experimentation could have prevented a huge problem.
It was my first visit to a very rural Virginia property,

Pennsylvania sedge forms carpets in open woodlands and could do the same in some areas where turf currently exists.

and my job was to design native gardens around the house and create a natural-areas plan for the extensive woods and fields on the property. A forester had previously advised the clients to extensively thin the trees in a large area of woodland, and they had already completed that task. This seemed a good idea. Red maples formed a monoculture throughout most of the woodland, and thinning them should provide openings for the colonization of late-successional species like oak and hickory. This would diversify as well as upgrade the species composition of the forest.

That all makes sense, except the site was poised to move in another direction. The soil, it turned out, was laced with seeds of black locust (*Robinia pseudoacacia*). Some oaks and hickories appeared, as the forester had hoped, but for every one of these, hundreds of black locust trees sprang up. This tree, although a North American native, is considered an invasive species in that part of the country by ecologists. In any event, it

is a straggling, weedy tree and would be considered a downgrade from the previously existing red maples by pretty much anyone. Had the forester tested the thinning and its aftereffects on a small plot, the property owners could have saved thousands of dollars and avoided a process that degraded instead of improved their woodland.

I believe that such a testing process should become standard practice when any kind of disturbance is being planned over a large area. If you plan to convert your lawn to meadow, scrape a section of turf, disturb the soil to activate the seed bank, and observe the response. Before clearing the invasive shrubs and vines in the woods, remove a few patches and you will get a very useful glimpse into the future *before* you've committed to expensive or time-consuming actions.

As we consider how to direct that trajectory away from the undesirable and toward the desirable, preliminary landscape trials can help provide those answers under real conditions and in real time. And in the case of the maple forest thinning project, it could have prevented a landscape disaster. Also pertinent is the question "What plants would I like to add beyond those that will naturally occur?" Here again, small-scale site testing of the plants on your wish list, particularly those you are relying on for extensive colonization, should indicate fairly accurately whether those plants will provide you with a diverse and verdant landscape or eat up your resources for naught.

Considering the site's past ecological trajectory helps to determine its inherent potential and your ability to favorably influence that potential *before* you decided what to plant. Essentially, you reversed the sequence by which landscapes are usually planned and created. You are in newly explored territory, and unfortunately there is no all-inclusive guide for navigating that territory, including this book. As with our sedge patch trial, trying out new plants and procedures at the small scale before expanding into the wider landscape may prove to be your most reliable resource for converting a long-shot idea into a pretty good bet.

CREATING
WOODLANDS

Woodland is the type of flora toward which succession trends in much of North America, throughout most of the area east of the Mississippi River, in the Pacific Northwest, and even in the Rocky Mountain region. Consequently, if you are standing in an open field, lawn, or even a flower border in one of these areas and ask our perennial question, "If I do nothing, what will happen?" your answer will be "A forest will develop." You might then think that establishing a woodland where one does not currently exist should be simple, a mere matter of standing back and letting things happen.

In fact, considerable management is involved in making sure you get the kind of woodland you want. The process moves forward spontaneously, but your participation is crucial to achieving a desirable result, and this may require a good deal of finesse. In particular, establishing the ground layer in a woodland is more difficult than, for example, establishing a meadow. That's because, unlike a meadow, a woodland ground layer is difficult to establish by seeding, and the woodland cannot easily be managed by mowing.

Consider first the issue of mowing. This is the principal means of managing a meadow, and in those circumstances it is a simple task. Cutting plants at certain heights and at certain seasons is just as effective in a woodland, but unfortunately woodland areas are usually too congested with trees and rocks to negotiate with mowers. Cutting in a woodland, as I will detail later, requires the use of slower and more laborious tools.

The issue of starting from seed is the more significant one. Here again, a woodland is very different from a meadow. Most meadow seeds can remain viable through a considerable period of dormancy; in some cases, the seeds will preserve their viability for years. In contrast, the seed of most woodland herbaceous species has a relatively brief period of viability. That is, these seeds die relatively soon after they ripen. By the time the woodlanders' seed has been harvested, cleaned, stored, and then shipped to you, chances are good that it won't germinate anymore. In addition, many woodland herbaceous species have very exacting requirements in terms of the conditions they must be exposed to before they will germinate. Even after they germinate, many woodland plants take several years to grow to a size that will make a visual impact in the landscape. Native lilies and trilliums,

Eastern bottlebrush grass and hardy ageratum, found in lightly shaded upland woodlands throughout much of eastern North America, are two of the few forest herbs that can be grown from purchased seed directly sown into the landscape.

for example, may take five years to reach blooming size, even in the ideal conditions of a nursery, and longer in a back yard. The solution is to let these plants sow their own seed, in a process described later in this chapter.

There are a few exceptions to the "don't plant woodland herbs from seed" rule. I've had success with direct seeding golden alexanders (*Zizia aurea*), eastern bottlebrush grass (*Hystrix patula*), and downy wood mint (*Blephilia ciliata*). As with the sedge trials I described, our firm continues to look for species and processes that can make seeding the woods a more viable approach.

Another characteristic complicates working in a woodland, and this is the typical presence of remnant native species. Generally, mat-forming European grasses dominate open fields, so establishing a meadow commonly begins with the elimination of all or most existing vegetation, a wiping clean of the slate. Woodlands, however, usually have at least some native species in the ground layer. It's rare to walk through an eastern forest without seeing white wood aster, wreath goldenrod, or some species of native fern or sedge. That is a good thing, of course, but it complicates weed control if you choose to preserve the natives. And here applying procedures that exploit the differences between desired and undesired vegetation again comes into play.

Taken together, all of these conditions mean that a program for planting and managing a woodland must be highly targeted and nuanced.

WORKING IN AN EXISTING WOODLAND

If your garden site is wooded, take some time to examine and understand what you find there before making any changes. This will help you to apply the right techniques in an appropriate manner when you undertake the management and enrichment of the landscape.

WOODLAND ZONES
An existing woodland has several different zones—the interior, the edge, and canopy gaps—each of which requires a distinct style of planting and management. The zones can be recorded on a plan or simply identified in the field.

Interior
The interior zone may experience fewer invasions from exotic plants because the light level is reduced. However, at least three particularly troublesome invasive species— Japanese barberry (*Berberis thunbergii*), garlic mustard (*Alliaria petiolata*), and lesser celandine (*Ficaria verna*)—do thrive in woodland interiors. The level of light that penetrates the tree canopy depends on the species present and how widely they are spaced.

Oaks, for instance, produce a canopy that is less dense and opaque than that of maples. Some tree species leaf out later in the spring than others, allowing more light to reach the woodland floor during that season, which can be beneficial to the growth of spring-blooming woodland wildflowers. This can also be a disadvantage, however. A study

performed by Lindsay Dreiss at the University of Connecticut found that woods dominated by late-leafing trees such as white ash (*Fraxinus americana*) and black locust tended to host more invasive species than woods dominated by early-leafing trees such as maples.

The spacing of the trees also helps to determine the light, as trees that are more distantly spaced tend to produce a less dense canopy. Aspect also plays a role: a woodland interior on a south-facing slope will experience more intense light than one on a north-facing slope.

Whether you decide to decrease light by planting additional trees or increase it by thinning or creating gaps should partly depend on the composition of the ground layer. Increased light will increase ground-layer growth. If the existing vegetation there is largely native, that is a good thing. If it's largely invasive exotic species, it is not.

Golden ragwort is one of many woodland wildflowers best installed as a live plant because it spreads primarily via rhizomes, making it a useful weed-suppressant.

Edge

The edge zone is usually a symptom of human activity and/or disturbance. For example, a road or driveway cutting through forest creates edges, as does the meeting of lawn and woodland. Nature's response to the creation of an edge is to seal it. Vines are commonly the sealant, as they grow up into the trees and create a curtain of foliage along the edge. This can be unsightly and detrimental to the affected trees. Also many of the species that

The planted ground layer in this managed woodland is denser than that found in a naturally occurring native woodlands, but this density is intentional in order to reduce opportunities for weed invasion.

take advantage of this opportunity are invasive, such as Oriental bittersweet. Wild grape (*Vitis* species) is a native that also performs this function, but it too can have the same undesirable effects. The vines do have some ecological benefit, as they protect the plants inside the edge from the sudden exposure to excessive light and desiccating winds.

A more aesthetically pleasing way to seal an edge is to proactively plant more attractive and less aggressive native vines, including Virginia creeper (*Parthenocissus quinque-folia*) or American wisteria (*Wisteria frutescens*). Alternatively, you can seal the edge with shrubs and trees such as eastern redcedar, American holly, sassafras, viburnums, and sumacs. All of these plants have dense foliage and are either evergreen or leaf out early, characteristics that make them highly effective for blocking light.

One disadvantage to sealing the edge is that you're likely to block the view into the woods. You can avoid this by leaving gaps between shrubs to direct the view into particularly attractive spots.

Gaps

The death and/or falling of trees create gaps in the tree canopy, areas with more sunlight within the woodland interior. The sunlight is rarely full strength except at midday, as the gaps are rarely big enough that the surrounding trees don't cast some shade. The higher level of sunlight within the gap fosters stronger growth of shrubs and herbaceous plants there, which can be both an opportunity and a problem. A gap is an opportunity to cultivate more plants and more diversity of plants in the ground layer, but it will also enhance the growth of any invasive plants that have found their way into the opening.

You may, for example, wish to preserve or even expand a gap if the natural ground layer there offers a strong growth of natives, or you may wish to close it by planting new trees if the gap is populated with invasive species. You can even create a gap by removing trees in an area with a favorable ground layer. Obviously, you shouldn't do this in an area infested with weeds and invasive plants. The stronger sunlight in a gap also promotes enhanced flowering, so they are a good place to locate plants that you want to produce seed.

WOODLAND LAYERS

While analyzing the woodland zones, you should also examine the vertical layers within the woodland to determine which, if any, are missing or sparse and, if so, why. Have browsing white-tailed deer impoverished whole layers of vegetation? Has the native vegetation been displaced by invasive species? Has human activity removed whole layers?

The four layers to look for are:

- canopy tree layer, which is the topmost layer, the interlacing of branches and foliage that shades all below them;

- understory tree layer, which consists of both permanent understory trees like serviceberry, pawpaw, and dogwood (*Cornus* species), as well as saplings of future canopy trees that occupy the space only temporarily;

- shrub layer, which consists of the shrubs that flourish under and among woodland trees; and

- ground layer, the lowermost layer consisting of herbaceous or prostrate woody plants on the woodland floor.

If any of these layers are lacking, you will have to replace them. You should also consider whether the canopy consists of high-quality native trees such as oaks and hickories or invasive species such as Norway maple (*Acer platanoides*) and black locust, the latter of which is native to North America but frequently becomes invasive outside its native range in the Appalachians and the mountains of Missouri and Arkansas. If the canopy consists of undesirable species, you may wish to gradually remove trees from the existing canopy and replace them with more desirable species or allow naturally recruited native trees to fill in.

You may also wish to plant to increase tree species diversity. Or you may wish to thin the woodland canopy, which is commonly the advice of any forester brought in as a consultant. Foresters, though, focus solely on the timber value of the trees and do not typically

Allelopathy is the phenomenon by which some plants release chemicals through their roots, foliage, and other parts that in turn suppress growth of other species. Eastern hayscented fern (*Dennstaedtia punctilobula*) kills or suppresses tree saplings and small shrubs as it expands, forming large monocultures.

consider the effect of their actions on the understory or ground layer, even though logging will have a major impact on these layers. If you consider thinning, keep in mind that increasing the light level inside the woodland interior will promote the growth of whatever is there, whether native or invasive. You may even wish to make the thinning selective, focusing this action on areas that tend to have more native vegetation on the ground layer. A cautious approach of thinning test areas first to observe the results of this action before applying it more widely can help you to avoid creating problems.

When dealing with an impoverished understory tree layer, you can plant appropriate species to make up the deficiency. Understory trees, incidentally, can be planted quite close to canopy trees as the two occupy different ecological niches and thus do not compete. This degree of spacing can even be desirable. An important function of understory trees is that they provide vertical bridges of cover, required by some bird species, between the shrub layer and the canopy.

A complete or near complete lack of an understory is, almost invariably, a symptom of excessive deer browsing. In such a situation, fencing off patches within the woods will allow the understory to replenish itself there and provide a seed source to populate adjacent areas if the deer herd is eventually reduced. As previously discussed, any trees that you plant in deer-infested terrain should be tall enough that the deer cannot reach the lowest branches with their browsing and should be protected against buck rub in autumn and early winter.

The shrub layer is often accentuated in disturbed woodlands. In a healthy woodland, native shrubs are usually scattered and sparse, whereas in a disturbed woodland invasive plants commonly become rampant and thicket forming, a condition incidentally in which tick-bearing white-footed mice proliferate. Japanese barberry is especially hospitable to mice. Recent research has shown this, although it has been common knowledge

The dappled shade on a wooded suburban property enables a diverse mixture of cardinalflower, sweetscented joe pye weed (*Eutrochium purpureum*), and woodland sunflower. The joe pye weed and woodland sunflower are long-lived species that will likely persist over time. The cardinalflower, a short-lived herb, will likely fade out of the composition unless intentional disturbance is applied to allow its seeds to germinate.

among hunters for many years. Reduce the invasive shrubs in the woods, and you will reduce the risk of Lyme disease.

Likewise, the lack of a native shrub layer may be a symptom of intense deer pressure. Other than a program of applying deer repellants, there is no easy solution to this problem, although it will help to avoid planting overly lush shrubs that were spoon-fed in the nursery. It seems that deer prefer the succulent growth that is present from intensively grown plants. I have seen nursery-grown mountain laurel (*Kalmia latifolia*) eaten when planted next to wild ones that were left untouched. By no means is this a panacea, but by using plants that have not been force fed for rapid growth, in combination with a repellant program, you may be able to keep your shrubs off the radar of these hungry herbivores.

Herbaceous plants in the ground layer typically grow in isolated patches or scattered sparsely, with much of the area covered by exposed leaf litter. Ideally, you should mimic this in your planting. However, where there is pressure from invasive species, you may need to plant a denser, more weed-suppressive ground layer. In essence, you will be creating an unnatural condition to keep things natural.

Many of the ground-layer plants are spring ephemerals that make most of their growth before the trees leaf out. In a woodland threatened with weed pressure, however, you may wish to ratchet up the presence of ferns and other late-emerging plants that can provide weed suppression beyond the spring season. This may also appear unnatural but often necessary to partially right the unnatural lack of balance that invasive exotics have produced in woodlands. Also these late species can generally coexist with spring ephemerals, as they occupy different niches in time.

In adding to the ground layer you should plant where there is adequate light to support the plants' growth. Visually prominent areas can also be selected to maximize the ability to enjoy the results of your efforts.

You should also consider existing tree species in locating ground-layer plantings. The filtered sunlight found under oaks and hickories is best for a diverse ground layer because it promotes the growth of a wide range of species. The dry, dense shade found under maples is less favorable, and the extremely dry shade found under American beech (*Fagus grandifolia*) is even less so. Very little grows in the ground layer of a beech-dominated forest. Beech forests, in fact, tend to be the least-invaded woodland type due to droughty soils conditions and the allelopathic effect of the leaves, which release chemicals toxic to other plants as they decompose on the forest floor.

MANAGING THE WOODLAND

A diverse herbaceous ground layer is impossible to establish by purchasing and planting seed, and seed distributed by the plants themselves is much more likely to germinate and produce offspring. Consequently, in the woods installing live plants is the way to go.

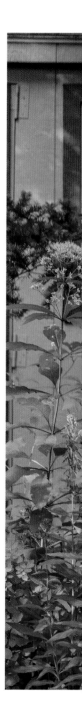

However, rarely is there a budget sufficient to fill an entire woodland all at once. Fortunately, with a fairly modest expenditure you can set the stage for the ground layer to plant itself. This is accomplished by creating mother colonies of mixed plants at suitable spots and then depending on them to self-proliferate. You'll achieve the greatest impact with this technique if you include ground-layer plants with a full range of proliferation strategies, from gravity-, wildlife- and wind-dispersed seeds to those that spread by rhizomes

MOTHER COLONY EXPANSION ZONES

Year 1 of management

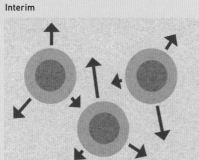

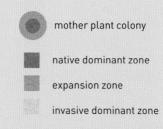

mother plant colony

native dominant zone

expansion zone

invasive dominant zone

Interim

End goal

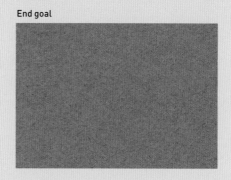

This diagram illustrates the steps involved in using mother colony expansion zones to establish and foster the expansion of herbaceous mother colonies in the woods. Making use of this incremental strategy also protects gardeners from the short-term but potentially huge burden of weed management incurred if they disturb the whole woodland floor by attempting to plant it all at once.

As shown in "year 1 of management," plant or locate existing dense patches of native herbs (native-dominant zones). Remove invasive species from these areas and continue weed control until the patch becomes thick and plants are producing seed.

Once the patch is dense, create an expansion zone around the mother colony by removing invasive species and lightly disturbing the soil to encourage germination of seeds from plants in the adjacent mother colony. Select for native species in the initial post-disturbance flush and continue to control weeds as in the inner native dominant zone. The most extensive weed control

efforts will occur in the expansion zone, as the inner zone will now be relatively dense and weed suppressive.

Meanwhile, invasive growth in the invasive dominant zone is periodically mowed or cut back during the growing season. Time the mowing to diminish resources of the plant and suppress seed formation.

As shown in the middle figure, over time the actively suppressed zone of invasive vegetation decreases in size, colonized by the native vegetation expanding outward from the mother colony. Once native species become dominant in the expansion zone, create a new one and repeat the management process. As the mother colonies are expanding, the invasive-dominant zone is becoming smaller, the plants are getting weaker, and no weed seeds are being produced. Conversely, seeds of native species have continually been distributed, giving them a greater presence in the soil seed bank.

Finally, as shown as the "end goal," the entire patch has become a native-dominant zone.

or stolons. Once the colonies are planted, you can then implement a program to increase the effectiveness and speed with which they expand into the surrounding landscape and eventually colonize the entire woodland area.

Many woodland herbaceous species, including spotted geranium, sedges, and ferns, are often lower growing than the weeds they commonly associate with, like Japanese barberry or Tatarian honeysuckle. This contrast in height affords an exploit-the-difference opportunity for inclusion in a management program. Cutting at a height immediately above

Weeds have been removed from an existing patch of Christmas fern, beginning the establishment of a mother colony.

the desired species damages only the undesired ones and has the predictable effect of weakening the weeds and putting them in an uncompetitive condition. Of course, tree saplings and emerging shrubs that exceed the specified cutting height should be marked before the cutting takes place to ensure they are preserved.

Although this strategy can be applied in almost any landscape setting, we most commonly employ it in woodlands, where it is usually performed with string-trimmers or hand-held brush cutters. Where the trees are widely spaced and the topography relatively smooth, it can be done with a field mower that allows for the blades to be set high. Perhaps as this style of management becomes more common, equipment manufacturers will design mowers nimble enough to negotiate the trees and uneven topography often found in the woods and with blades that are easy to raise and lower.

As strange as this might sound in the context of natural design, you could call this hedge trimming the woods. But here instead of cutting plants to preserve the status quo, you are cutting plants to influence change. When the desired change occurs and the favored vegetation becomes dominant, you can cease performing the activity. In effect, you are performing work now to decrease work in the future. This is the opposite of landscaping in one fell swoop and a hallmark of our process-based, cause-and-effect approach.

Fallen leaves can present a bit of a dilemma to the woodland gardener. They suppress the growth of many woodland wildflowers, which often grow on hummocks or fallen, rotting logs from which the wind strips the leaves. Yet the decaying leaves are an essential element of the topmost layer of the woodland soil. One way to increase the opportunities for wildflowers without damaging the soil is to shred the leaves with a mulching mower, shredding them in place if possible or, if there are too many obstacles to mowing, by

A standing dead tree (snag) was intentionally left to serve as a food source for insects that feed on decaying wood and are a crucial part of nutrient cycling. The tree's cavity could also act as a nesting site for birds and small mammals. Aside from its habitat value, the snag serves as a testament to the site's prior vegetation.

blowing the leaves into a pile, shredding them, and then blowing them back over the woodland floor. Before mowing, however, you should mark with stakes any existing wildflowers, saplings, and shrubs that you want to avoid damaging.

Physical structure in the woodland creates hiding and nesting places for wildlife and creates small microclimates that increase plant diversity. Logs, boulders, and walls all create such structure and should be preserved, if present, or even placed within the woodland if it is lacking structure.

Dead and downed wood, in particular, plays an important role in the life of the woodland. Snags—dead but standing trees—are nesting sites for a variety of wildlife. Fallen sticks and twigs create fertile spots where seedlings sprout and nestle. Of course, as the wood decays, it plays an important role in nutrient recycling.

For all these reasons, the deadwood should not be removed. If the logs are big and obtrusive, they can be cut into pieces and laid flat on the ground, or even sliced into rounds and used as short-lived stepping-stones. Or branches and logs can be heaped into brush piles, which serve as retreats for small mammals, reptiles, and amphibians.

At the Mt. Cuba Center in Hockessin, Delaware, a fallen tree has been left to decompose, returning nutrients to the soil and providing a spot for plants such as this Christmas fern to become established.

WOODLAND SOIL

Woodlands are composed of layers aboveground, but the same is also true below the surface. Typically, a woodland soil consists of the following horizons or layers in descending order:

L horizon, the uppermost layer, is composed largely of undecomposed leaves, bark, and twigs;

F horizon, the next layer down, consists of partly decomposed materials formerly in the L horizon;

H horizon, the deepest organic layer, consists of fully decomposed organic material; and

mineral soil, derived from the underlying bedrock, does not contain significant amounts of decomposing organic material.

MAN'S ROLE IN NATURE

I remember an occasion years ago involving a client's property on which a huge old European beech (*Fagus sylvatica*) had died. I suggested that the lower portion of the main trunk be left standing as a snag for wildlife. The arborist cut the trunk back to a height of about 12 feet, and I directed him in making the final cut in a way that it resembled a natural storm break. I asked him to change the angle this way and that, until a feeling of ridiculousness suddenly overtook me. If a beaver had cut the tree, I would be glorifying it as a natural occurrence, not hiding it. I am an animal, too. Why do I feel so compelled to hide my role in this tree's journey? So I told the arborist just to make a flat cut and move on. I suspect that the beauty of our handsome wildwood sculpture wasn't diminished in the least. The beaver and I are equally a part of the ever-changing milieu that is nature.

Tilling such soil, mixing it into a homogenous whole, creates a condition inimical to woodland plants and more hospitable to invasive weeds. Therefore, if you wish to apply compost to your woodland, perhaps because the top layer has been eroded away or digested by earthworms, use composted leaves and distribute it in a layer across the surface. Do not dig it in.

THE BIG PICTURE

When working with established woodland, your initial goal should be to transfer the vegetation from exotic, invasive species to a flora dominated by native ones. Once this has been accomplished, you may take as your subsequent goal to translate the vegetation from generalist native species to more specialist natives. This second process will eventually proceed on its own, particularly if you are controlling invasive species, but the progress is likely to be very slow. If no seed bank remains of the specialists, which is often the case because seeds of woodland herbs appear to have short viability, it may not happen at all.

You can accelerate or initiate this conversion by introducing container-grown plants of the specialist species, and use the techniques I described above to foster their expansion. If you are so fortunate as to already have the specialists in place, keep them there by avoiding disturbance.

INITIATING NEW WOODLAND

In titling this section, why did I choose *initiating* instead of *planting*? The reason is that in most of North America if you do nothing, trees will grow. You don't have to plant a woodland so much as set this process of regeneration in motion. Yes, you are likely to do some planting yourself, but once you initiate regeneration, you will usually find that nature supplies many of the required plants for free.

CONSIDERATIONS BEFORE PLANTING

This process begins with decisions about what trees to plant. Your choices will have ramifications in the type of habitat they create and the type of woodland that will result. For example, if a diverse ground layer is your goal, then American beech is not the best tree to plant as few herbaceous plants grow in its dense, dry shade. However, if low maintenance tops diversity on your priority list, then beech may be a very good choice.

The degree of shade is also important to consider. Although the site may be sunny when you start your planting, you should keep in mind what result the process will produce. What degree of shade do you desire? Do you want fairly open woods supporting a vigorous and diverse ground layer? Or do you want deeply shaded woods more resistant to infestation by invasive species? Do you want late-leafing trees such as ash (*Fraxinus* species) that will foster the growth of spring ephemerals, or do you want early-leafing ones such as maple that will help to exclude invasive species?

Again, remember that the trees you plant should be tall enough that the lower branches are out of the reach of browsing deer, and that you must protect them against buck rub in autumn and early winter.

PLANTING FOR NATURAL SUCCESSION

To create a successful woodland, understanding the process of succession is not an option but a necessity. Young woodlands are usually dominated by sun-loving pioneer species like birch (*Betula* species) or aspen (*Populus* species). The shade they eventually create fosters long-lived, mature forest species like oak and hickory.

A fine illustration of how this aspect of succession can be used to create a woodland is found at Mt. Cuba Center, a former estate turned botanical garden in Hockessin, Delaware.

EARTHWORMS

Most gardeners were raised to look upon these creatures as an unmitigated good. The fact is, though, that earthworms are not native to the northern part of North America; they were extirpated from this region during the last ice age. If there are earthworms in a northern woodland, they are invasive species causing considerable harm by rapaciously consuming the layer of litter and duff that naturally carpets the forest floor and that is essential to the life cycle of many native ground-layer plants. Essentially, their action homogenizes the soil, destroying the naturally layered character of a woodland soil.

You can easily identify their presence by examining the soil, which will have been turned into a near weightless compendium of hollow-appearing shells. Fortunately, earthworms tend not to be an issue in highly acidic forests, like coastal pine barrens or longleaf pine communities, as they do not tolerate the low soil pH found there.

Not much can be done to eliminate earthworms where they are present, but at the least you should avoid introducing them to woodland areas not already infested. You can also make a point of regularly replacing the topmost organic layer to mitigate the effects of earthworms. Also, plant the most generalist, aggressive woodland species such as white wood aster or Christmas fern in your ground layer. These may take a long time to establish in these conditions and may require irrigation initially, but as they mature, such pioneers seem to change the soil dynamics so that other, less aggressive species can follow in their wake.

Mt. Cuba focuses on the native vegetation of the local Appalachian Piedmont. Beginning in 1983, Mrs. Lammot du Pont Copeland and the center's first director, Richard Lighty, began returning some 550 acres of mostly open fields to the native woodland garden that exists today. This remarkable transition was achieved in so short a time by planting the fields with fast-growing, native tuliptree (*Liriodendron tulipifera*), an early-successional species. Late-successional trees such as oaks, hickories, and maples were either planted among the tuliptrees once they provided shade and a suitable habitat or were naturally recruited. Without learning the history, few visitors would realize that many of the trees defining this fully mature cathedral-like woodland were actually planted, and how ecological succession played such an important role in sequencing that planting process.

Woodland habitats are shady, but a site at the beginning stages of reforestation is not; nor are the conditions there favorable, initially, for the establishment of a woodland ground layer. Indeed, until the young trees grow up to produce an overarching canopy, woodland herbaceous species will not grow underneath and around them. For this reason, the initial groundcover on the floor of a woodland restoration should be one of sun-loving meadow plants with an intermingling of more shade-tolerant species adapted to filtered light conditions that the trees will provide over time. The meadow plants provide a nurse crop of sorts in the early, sunny stage of the restoration, covering the soil and suppressing weeds. It is important in this scenario to plant trees taller than the meadow vegetation so as to allow them to be easily located and maintained during the establishment period.

When meadow is the long-term goal, you mow or burn the area annually to prevent tree encroachment. In this effort, however, tree encroachment is a good thing and can supplement your planting efforts. If this is desired, you should allow this process to unfold by not cutting the meadow and limiting your management efforts to spot spraying or cutting of invasive species.

Eventually, of course, a condition of deep woods develops. As I have already mentioned, many of the species that populate the ground layer of a woodland interior produce short-lived seed, and planting them at the outset of the restoration effort would be pointless. Instead, delay planting the woodland ground layer, either from seed or more commonly with plants, until the young trees have matured sufficiently to produce a substantial canopy. At this time, as the deep woods condition emerges, the gardener may also switch to the management techniques described previously in this chapter.

If you do choose to use meadow as a nurse crop for your woodland restoration, mowing will be a necessary part of the initial management. To facilitate this, instead of setting out trees in a random pattern, plant them in a grid that creates linear paths in two directions so that the mower can pass up and down.

PIONEER TREES TO FOSTER MATURE FOREST SPECIES

A certain degree of shade is essential to create favorable conditions for the growth of late-stage succession trees. To commence the woodland restoration program, you can follow the example of the Mt. Cuba Center and begin your planting with fast-growing pioneer species like tuliptree or birch, and then plant the late-stage species like blackgum, oaks, hickory, or beech in their shade.

Meadow was planted as a nurse crop for this reforestation planting. The white tubes contain small saplings of native woodland tree species.

Three years later, the trees are growing out of their tree tubes while the meadow remains only in the sunnier spaces between the trees. Trees were excluded from the drainage swale in the foreground, and meadow will be the long-term composition here. Yearly cutting will prevent tree encroachment and preserve the swale at the meadow stage of succession.

As the tree cover phases in, the meadow phases out, as per plan.

Over time, the late-stage trees will displace the pioneer species. In small plantings, this process can be hastened by actively removing the pioneers. An example of this technique might be a situation in which you wish to screen the view of something tall and unattractive such as a telephone pole. The goal is to create a fast-maturing screen, but also a durable one that isn't prone to storm damage. These characteristics are usually mutually exclusive, as most trees are either one or the other. Both can be had, however, by planting initially a fast-growing (but damage-prone) species such as tuliptree or American basswood (*Tilia americana*) and then, when that has attained some height, planting beneath it a more durable (less damage-prone) oak or maple. Continually removing the lower branches of fast-growing species as the late-stage tree follows behind will give the latter the room it needs to grow. When the more durable tree matures sufficiently to provide the desired screen, the fast-growing tree can be removed. In this way, you use the mechanism of succession to solve your landscape problem quickly but in a manner that will also function well over the long term.

NATURAL RECRUITMENT

If you live outside of the prairie states and the arid West and don't mow, trees grow. Because of this, woodland creation offers great opportunities for incorporating natural recruitment into your planting program. I recall driving by a beautiful expansive grassland in the coastal plain of Maryland. When I stopped to walk the field and take a closer look, I encountered a diverse array of native grasses, flowers, shrubs, and small pioneer trees. The field was largely wet, which explained the extensive presence of native species. Wet and dry extremes tend to host naturally occurring native species most commonly, particularly in open fields. I noticed, however, that small sweetgum (*Liquidambar styraciflua*) trees had been planted among the naturally occurring vegetation. One of those sweetgums was immediately adjacent to a naturally occurring one, which was almost twice the planted tree's size. As one would not likely plant a sweetgum tree a foot away from an existing sweetgum, I assumed that the naturally recruited one must have been too small to notice when the new tree was planted. Yet this tiny tree had overtaken the larger planted one.

The first lesson to learn from this is that a naturally recruited plant will almost always outperform a planted one of the same species. The second lesson is even more important. If you understand the successional process of a wet meadow in the eastern coastal plain, you will know that sweetgum trees almost always come in by the thousands. Why plant them then? Why not plant a wet meadow tree that is less likely to occur on its own like blackgum (*Nyssa sylvatica*) or American sycamore (*Platanus occidentalis*)? Or put resources into uprooting the few invasive multiflora rose shrubs that I also observed on my walk? When I visited they were few and far between, but if allowed to persist, they won't stay that way.

While I am sure the tree planting effort was well-meaning, it was clearly redundant and really didn't accomplish much. How could this have been avoided? By asking the question, "If I do nothing, what will happen?"

The owners of this property in Hastings-on-Hudson, New York, hired our firm to complement their Arts and Crafts-style home with a landscape that reflected both the ecological character of the region and the historic character of house.

Passing through this gate and descending the stone steps leads the visitor through a series of sitting areas and viewing points.

top Chairs on a small patch of lawn face a short meadow garden that foreshadows the wilder plantings on the opposite side of the property.

bottom and left This pond and waterfall were designed to appear as though they were naturally occurring.

POSTSCRIPT

LIVING IN IT

Many would consider landscape painting to be the closest art form to garden design. I think that in one important regard, music composition is closer. A landscape painting is a static thing that hangs on the wall, whereas a musical piece unfolds over time, as does a process-based natural garden.

Natural design and contemporary music also share an embrace of the element of chance. The late John Cage, an iconic twentieth-century composer, introduced chance into the heart of his compositions. He would select notes or sounds based on a roll of I Ching dice or the sound of a truck that happened to pass by his window as he composed. In some cases, the musicians could play a series of notes that Cage composed, but in any order they wanted. Here he was setting up a framework in which the musicians must remain—the notes he selected—but he left part of the composition to chance—the order in which they were played. Although my musical tastes often veer to the fringes, Cage's approach used to leave me cold. "If you want to write music, write music," I thought.

An experience I had while working on a new musical composition changed that perception, however. I was creating a horn part to go with a particular flute melody. I recorded the melody and two alternative horn parts. Each horn part was written to go with the melody, but not with each other. For a lark, I decided to play all three together. As I expected, the two horn parts mostly conflicted and the combination sounded horrible, except in two places where the horns created phrases that were rhythmically and harmonically wonderful and probably more interesting than anything I could have written intentionally. I kept those sections in the piece and decided to expand the ensemble to two horns and a flute as a result.

This experience gave me a better understanding of what John Cage was really doing. Elements of his music were in fact random, but he often determined which ones to select and the framework in which they would occur. The result was part random occurrence and part composition, as was the case in my piece. I had selected only two of the many sounds that resulted from my random horn pairings. In addition, the melody provided a framework in which those random elements would occur. While the horns were not directly relating to each other in my overlaid recordings, they were both relating to the same melody and were thus indirectly connected.

I realized that Cage's approach to music composition had tremendous relevance for me as a landscape designer. Much of what occurred in my home landscape was random. Like Cage, however, I was selective about which occurrences I would allow to play out. But there was also an ecological framework that influenced these occurrences. Where a wind-blown seed landed was random. Whether it germinated was not, because soil conditions and competition from other plants would determine that. If it did germinate, it was up to me whether the plant remained. I still marvel at how often the confluence of these events, as in my musical piece, resulted in plant compositions that likely exceeded my design talents.

MY HOUSE

I had always intended to create a master plan for my own property—next month. But next month never came, and my tiny ⅓-acre spread on the outskirts of Philadelphia advanced without the benefit of a plan. That doesn't mean I exerted no influence. I planted. Some plants found their way to my garden because I put them there intentionally, some were castoffs from landscape projects, and some arrived on their own.

I learned from them all over the twenty-eight years that I've lived here. From the plants I selected intentionally, I learned to hone my "match the plant to the habitat" skills. Also I learned how to make a wild garden look attractive on a small property.

From the castoffs, I learned that serendipitous decisions can sometimes work out and sometimes not. After renovating a woodland garden for a client, I ended up with clumps of an unidentified exotic fern. The client wanted them out, as they were "running amok" in his garden. I thought "How bad could a fern be?" and planted them at my house, where they promptly ran amok. I was so annoyed with myself for needlessly planting them that I refused to spend time pulling them out by the roots. Instead, I quickly raced through the garden ripping off the fronds at the base to at least hold them at bay. After a few ripping sessions, they faded out of the picture. I realized that no plant can survive indefinitely when deprived of leaves and the ability to produce food, especially in the intense competition that existed in the ground layer of my garden. I extended that technique to other weeds and noticed their general abundance began to decrease after I instituted this approach. I realized that my new weeding process was no longer creating soil disturbance and stim-ulating weed seed germination. From that experience arose the genesis of the "cut, don't pull" weeding method that I have advocated ever since. This experience also contributed to my understanding that weeding doesn't need to succeed all in one session; alternatively, you can take actions disfavoring the weed over time, let the competing plants finish the job, and achieve the same effect with less effort and disruption.

But it was from the volunteers that showed up unannounced that I learned the most. Eastern daisy fleabane (*Erigeron annuus*), an herbaceous biennial, appeared shortly after I removed an inherited bed of English ivy. The plant has an attractive white flower, but it has a rank growth habit and is generally considered a weed. I was on the fence whether or not to remove it, but took the path of least resistance and let the fleabane remain. By the following year, it had disappeared anyway, competed out by the golden ragwort, spotted geranium, and other ground-layer herbs that I had planted. In thinking about this sequence,

I realized that of course the fleabane wouldn't last. It is an early-stage, short-lived disturbance species, and it was competing against the long-lived, site-adapted perennials that I had planted. Armed with this information, I could have asked, "If I do nothing, where will this composition go from here?" The answer would have been "a nice stand of golden ragwort and spotted geranium, *without* daisy fleabane." Accordingly, the best course of action was to do what I most enjoy doing in my garden: nothing.

Common blue violets (*Viola sororia*) showed up by the hundreds, but this was a no-brainer. Why would I remove a creeping groundcover with attractive heart-shaped leaves and flowers of white and blue? In a mulch bed those leaves are highly visible as they turn brown and ugly in the heat of summer, but in my mixed ground-layer composition they are covered by other foliage and out of view by that time. The violet flowers look great intermingled with the planted native woodland stonecrop, which has since migrated to various gaps and openings in the landscape. Other planted self-proliferating species include smooth Solomon's seal (*Polygonatum biflorum*), meadow-rue, tall phlox, seersucker sedge (*Carex plantaginea*), joe pye weed, hairy alumroot, narrowleaf blue-eyed grass (*Sisyrinchium angustifolium*), red columbine, Christmas fern, and bottlebrush buckeye. I've been observing how far these volunteers traveled from the mother plant, information that has helped me to determine which seeds need distribution assistance and which ones don't.

top Although daisy fleabane may be a weedy-looking plant, I did not feel compelled to remove it as it will be outcompeted over time.

bottom Why would I want to weed violets?

One year a storm broke off half of an old white oak in the rear corner of my yard. The fallen branches dislodged the existing ground-layer vegetation and disturbed the soil. The response was a new patch of sassafras trees and a lone Oriental bittersweet, a highly invasive exotic vine. If I did nothing, the bittersweet would bury the sassafras, so I removed it. Now every year I enjoy the bright yellow of an autumnal sassafras grove from my kitchen window.

Volunteer plants were coming and going according to their own proclivities and devices, and my role in that process was deciding which could stay and which would go: landscape editing of sorts. The next step in my learning process was to go beyond simply editing what naturally occurred and to investigate how I could proactively influence those occurrences.

I had come across a study performed at Rutgers University by Jean Marie Hartman and her colleagues that analyzed the vegetative responses to different treatments applied to small plots of bare ground. One plot received a mulch of raw wood chips, and a greater number of native trees sprang up there than in any of the other plots. Most interestingly,

the species that appeared were not the early-stage pioneer trees like the sassafras in my back yard, which one might expect to colonize, but late-stage species like oak or hickory that usually won't sprout until after pioneer tree species have established themselves and cast some shade.

This was of great interest to me as I wanted more shade trees around the stone terrace in my back yard. Hundred-year-old specimens of white oak, scarlet oak, and shagbark hickory flourished elsewhere on the property. Clearly, if I wanted to plant trees that were adapted to my property, these species would be good choices, as they had certainly withstood the test of time. But did I really need to plant them?

Instead, I elected to apply raw wood chips, and it worked. Numerous oak and hickory seedlings emerged from the chips, whereas almost none sprouted elsewhere on the property. The Rutgers researchers did not explain why such a thing should occur, but possibly the mimicking of the leaf and twig litter on a woodland floor by the wood chips had prompted acorns and hickory nuts to germinate. Whatever the reason for the effectiveness of this technique, it had allowed this area of my garden to skip over the pioneer

Golden ragwort and other more aggressive ground-layer plants did the job for me.

A huge branch fell from this oak and disturbed the pachysandra below. The response was a free sassafras grove.

tree stage of succession and move directly the later stage, where long-lived canopy tree species dominate.

Of course, these trees would not have grown there if no seed source was close by. Small mammals like squirrels and chipmunks distribute the seeds, and they don't travel more than about 30 yards from the mother tree to bury them. I did have the seed source, and I was overjoyed when my miniature oak-hickory forest appeared.

But there were more tree surprises in store. Elvis, my small Shetland sheepdog, had a favorite spot for performing his business. His frequent visits caused the spot to become muddy, and mud was then tracked into the house. For this very high-minded reason, I covered the area with small, round pea gravel and hoped the kitchen floor would henceforth remain clean. It did, but something else also happened. The area became carpeted with tiny seedlings of native black cherry (*Prunus serotina*) trees. I had no idea why this occurred. When I consulted an ornithologist friend, however, I learned that birds swallow the stones to use in grinding their food in their gizzards. While engaged in this activity, they are likely to pass seeds, which was why the cherries were concentrated there in such large numbers.

As we have noted, trees will grow pretty much anywhere that woodland vegetation is native if you don't weed them or mow them. Here, though, I was actually influencing what type of tree grew. I could disturb the unaltered mineral soil, as the fallen oak had done, and get sassafras. I could apply wood chips and get oaks and hickories. My placement of a thin layer of pea gravel produced black cherries. Not only could I influence which tree types colonized my property, by selectively locating placement of the material that fostered their establishment I could influence where they colonized. I could design with wood chips and gravel, right down to the arrangement.

The application of raw wood chips generated these oak and hickory saplings.

Recently I revisited a site in Allentown, Pennsylvania, where my firm had planted 12- to 14-foot tall northern red oak trees many years ago. While there, I realized that the trees were planted the same year as the oak-hickory-producing wood chip application in my yard. A rough calculation of both of their sizes revealed that my naturally recruited trees were bigger, even though the planted ones had had quite a head start.

Naturally recruited plants will almost always outperform planted ones, which is another reason to take advantage of the plants that are waiting in the wings or, more accurately, in the soil. Consider the itinerary of a nursery-grown tree traveling to a residential yard, let's say a northern red oak to be planted in suburban Philadelphia. The grower could

This naturally recruited grove was sited not by planting the trees but by deciding where to spread the wood chips.

be located in Tennessee, where a huge nursery industry ships all over the east coast.

First the northern red oak will be dug by a mechanical tree spade in the nursery and loaded onto a tractor trailer. It then begins its journey on a highway winding through Tennessee forests full of naturally occurring northern red oaks. Then it passes through the woodlands of Kentucky, where more northern red oaks occur. West Virginia, Virginia, and Maryland, northern red oaks all the way. The trucker may be bored and gazing out the side window by the time he hits Delaware and Pennsylvania. If he does, he'll probably see northern red oaks.

Finally the driver arrives at his destination, a landscape wholesale yard. Here irrigation towers will distribute water, some of which will end up on the tree and some will run off into the stormwater system. At some point, a landscape contractor arrives and uses a forklift to load the tree onto a truck, which carries the tree to its suburban Philadelphia destination. There, another fossil-fuel-driven machine is waiting to carry the tree to the back yard of the property, and with the help of a landscape crew, the tree is planted.

In my yard, a chipmunk scampered 30 yards and planted an acorn. And my northern red oaks are bigger than the Allentown trees I planted the same year.

Of course, when I design landscapes I still plant plenty of nursery-grown herbs, shrubs, and trees. But doesn't it make sense to also plan for and manage naturally recruited plants? After all, they cost even less than the most marked-down plant in the autumn blowout sale at your local garden center, and they are far more likely to survive.

Another very educational sequence of events in my garden involved not trees but the herbaceous plant cardinalflower, the plant with which this book began. I had removed the existing English ivy from a bank above my stone terrace, and shortly afterward planted a mix of native plants there, including cardinalflower, a short-lived, relatively ephemeral species. I expected it to be a temporary player that would yield to the more aggressive planted species; and it did. Not only did the original plants die, but no new seedlings emerged. This was expected because of the dense ground layer that had developed on the site.

A number of years passed and I saw no cardinalflowers in my yard, until one spindly plant emerged from between a grouping of stones. Hoping to bring the plant back into the fold, this time in a lower garden within the stone terrace, I picked a seed stalk and placed it on bare soil in a gap next to the flagstones. The next year a seedling sprouted from the gap, and the following year it flowered and went to seed. Hundreds of seedlings emerged in the cracks between the stones of the terrace the next year. The year after that, a few of these plants flowered, and the following year the terrace was a sea of red cardinalflower. A pair of hummingbirds hovered over the terrace all summer, and my main task was to remove enough of the lush, red-flowered growth to create paths and a place for my dining table.

What had happened here? The cardinalflower had perished from the bank above the terrace years ago. The bank is directly above the terrace, yet no seed had ever washed down to the terrace and plants had never emerged from it until I moved a seed stalk right next to the stone.

The answer, I believe, lies in the species' seed dispersal characteristics. The plant produces dust-like seeds that fall directly at the base of the plant and attach themselves to soil particles. This attachment prevents them from being carried by runoff during rains. Consequently, the seeds never moved from the bank to the terrace below. But when I laid a stalk down directly adjacent to the terrace, its seeds fell onto the stones. With no soil to

The cardinalflower drama plays out around my stone terrace. One seed stalk produced it all.

Phlox, hairy alumroot, woodland stonecrop, and other woodland plants native to the Northeast colonize my front yard from only a few planted individuals.

attach to, they washed across the terrace and into the joints, where they germinated. The migration wasn't finished, however. The cardinalflower continued its march, but only along the stone paths leading away from the terrace, where no soil existed to bind the seeds in place.

The takeaways here were numerous:

- Cardinalflower seeds won't germinate in a competitive situation. Once I placed them adjacent to the terrace, they found open space between the stones where no competition suppressed them, and they went to town.

- Cardinalflower seeds would have to be distributed by the gardener to colonize extensively, even if it is already growing uphill from the area you wish it to colonize.

- You can lead cardinalflower by placing stone ribbons, paths, or other hard surfaces through the landscape.

- Because of how they bind to the soil, cardinalflower seeds are likely to remain in place, even if sown in a drainage swale or other situation through which large volumes of water frequently pass.

- Other floodplain species, many of which have seeds with characteristics similar to those of cardinalflower (including most wetland sedge species and joe pye weed), can be treated as described here.

The lessons continued. Alongside one of the paths that led from the rear terrace to the front of the house, I had spread pea gravel on a narrow strip between the path and the house. It was placed there solely for utilitarian reasons, to provide a dry, mud-free surface in this little-viewed area for hose bibs and the like. To my surprise, hairy alumroots planted elsewhere in the yard seeded in—not just a few scattered plants but a perfectly solid cover, as good as any gardener could plant.

I had observed hairy alumroots seeding before, but rarely in the clay soils found throughout most of my property. Here the plants grew almost exclusively from gravel areas at the base of a wall or around the stone bench in the front yard. They also grew from a brick-encircled cement pad left over from an old construction effort. After they hopped from those spots to my gravel strip along the pathway, I finally got the message: alumroot seeds like to germinate in gravel. That makes sense, as the species grows on dry ledges in nature.

The opposite side of my hairy alumroot path consisted of a narrow strip of pachysandra that ended at my property line. I decided that being led down a path of alumroots would be a nice thing, so I removed the pachysandra and laid down some pea gravel. I was done. No plants needed. The hairy alumroots that colonized the other side of the path would do the planting for me. The first colonization was a happy accident. The second was planned, but only after I observed, analyzed, and acted on that happy accident.

The next spring I checked the gravel to see if there were alumroots germinating. There were, along with cardinalflower seedlings that had evidently continued their march from the rear terrace and down that same stone path. Now I have an alumroot-cardinalflower border along my path. Neither species was planted, but neither would have been there if I hadn't set the stage.

If I do nothing further, where will this composition head from here? Given that cardinalflower is a short-lived, noncompetitive perennial and that alumroots form dense long-lived carpets, the alumroots would shortly overwhelm the cardinalflower. However, a little applied disturbance, about thirty seconds of raking at the base of the cardinalflower plants, will stimulate germination of their seeds, and keep them coming back.

Do you see what's going on here? Plants I planted are moving around. Plants I didn't plant are moving in. Some of these volunteers I leave, some I remove, or at least cut off the top parts. A storm break rewards me with a sassafras grove after I edit the vegetative reaction. I apply specific material to foster specific trees, which results in an oak-hickory grove. These trees are now more than 40 feet tall and form the visual centerpiece of my

top The hairy alumroots above the wall were planted. The ones at the base naturally recruited, but only in the gravel of the adjacent terrace.

bottom After I placed gravel between the path and the wall, alumroots filled the space. Once the pachysandra on the other side of the path was removed, it was colonized by the alumroot as well.

entire back yard. Herbaceous plants are being led around the property as I move their seeds or create pathways for them to move on their own. This, certainly, is a very different way to experience a garden.

A give-and-take relationship like this doesn't happen in a garden that is completely arranged and static. You are the boss there, even though you may struggle at times to keep your employees under control. But neither is a connective experience likely to occur at the opposite end of the spectrum, during, say, a visit to a remote natural area. The vast and picturesque views at Yosemite National Park may be exhilarating, but I had no effect on this landscape and consequently no personal relationship with it. Conversely, I've experienced twenty-eight years of concrete and mutually beneficial interactions with my landscape, a partnership with nature.

The phrase "a partnership with nature" has been around a long time, and I never paid it much attention. From a practical standpoint, I really didn't know what it meant. Now I think I do.

Not everything I learned at my home is incorporated into every one of my projects. The tastes, interests, and functional requirements of my clients vary, and I have responded with gardens that range from completely wild to quite formal and everything in between. Actually, I enjoy the variation. When you loosen the reins and work collaboratively with the landscape, consulting and respecting what it wants to do, the experience moves to a whole new level.

One type of experience that has long been a part of garden design is to create a sense of mystery. A garden path winds through a tall border of shrubs, emerging suddenly to reveal a carefully placed sculpture. Let's be real, though. For the folks who live with that garden, who have traversed that path a hundred times, there is no mystery. Only new visitors will experience the desired effect.

left Thanks to the cardinalflower, this hummingbird is a constant visitor in late summer.

above I prefer a glass of wine and a good book to weeding.

But if an owner's walk reveals plants that have migrated from area to area, new plants that have appeared on their own, or a sea of woodland phlox, the results of a seed stalk they placed in a gap last year, there *is* mystery around the corner in their own back yard: the corner of time.

How do I spend the time that falls between these surprises and revelatory events? Instead of weeding, pruning, and endlessly rearranging the composition to achieve artistic perfection, I may just spend some quality time in the hammock or enjoy a glass of wine with my drinking partner.

METRIC CONVERSIONS

inches	cm
⅛	0.3
¼	0.6
½	1.3
¾	1.9
1	2.5
2	5.1
3	7.6
4	10
5	13
6	15
7	18
8	20
9	23
10	25

feet	m
1	0.3
2	0.6
3	0.9
4	1.2
5	1.5
6	1.8
7	2.1
8	2.4
9	2.7
10	3.0

acres	ha
⅓	0.12
1	0.40
5	2
10	4

ounces	g
1	28.3
5	141.7

RESOURCES FOR GARDEN ECOLOGISTS

ORGANIZATIONS

These are some of the many organizations that focus on ecological restoration, native plants, and native plant conservation and research.

American Horticultural Society: ahs.org

Botanical Research Institute of Texas: brit.org. Focus is worldwide.

Ecological Landscaping Association: ecolandscaping.org

Lady Bird Johnson Wildflower Society: wildflower.org

Native Plant Center, Westchester Community College: sunywcc.edu/about/npc

Natural Areas Association: naturalareas.org

Society for Ecological Restoration: ser.org. Chapters by state and/or region.

Wildflower Conservancy: wildflowerconservancy.org. Committed to conserving rare, threatened, and endangered species of wildflowers and the larger ecosystems coexisting with them.

JOURNALS

These periodicals can be useful sources of emerging research. Several of the publications are available as digital subscriptions (which tend to be less expensive). Some may be best accessed through an academic or specialty library, as the subscriptions are costly.

American Gardener, American Horticultural Society, published six times annually, subscription included in American Horticultural Society membership fee.

Bulletin of the Ecological Society of America, Ecological Society of America, published quarterly, available only online.

Ecological Application, Ecological Society of America, published eight times annually, peer reviewed.

Ecological Monographs, Ecological Society of America, published quarterly, peer reviewed.

Ecological Restoration, University of Wisconsin Press, published quarterly, peer reviewed, discounts available to Society for Ecological Restoration members.

Ecology, Ecological Society of America, published monthly, peer reviewed.

Frontiers in Ecology and the Environment, Ecological Society of America, published ten times per year, peer reviewed.

Grasslands, California Native Grasslands Association, published quarterly, subscription included with California Native Grasslands Association membership fee.

Invasive Plant Science and Management, Allen Press for the Society for Weed Science, published quarterly, peer reviewed.

Landscape Journal, Council of Educators in Landscape Architecture and University of Wisconsin Press, published biannually, peer reviewed.

Native Plant Journal, University of Wisconsin Press, published three times yearly, peer reviewed.

Natural Areas Journal, Natural Areas Association, published quarterly, peer reviewed, subscription included in Natural Areas Association membership fee.

Nature Conservancy Magazine, Nature Conservancy, published quarterly, subscription included in Nature Conservancy membership fee.

Restoration Ecology, Wiley Periodicals for the Society for Ecological Restoration, published eight times annually, peer reviewed.

Rhodora, New England Botanical Club, published quarterly, peer reviewed, subscription included in New England Botanical Club membership fee.

Urban Ecosystems, Springer for the Society for Urban Ecology, published quarterly, peer reviewed.

Wetlands Ecology and Management, Springer, published six times annually.

Wildflower Magazine, Lady Bird Johnson Wildflower Center, published quarterly, subscription included in Lady Bird Johnson Wildflower Center membership fee.

WEB RESOURCES

Biota of North America Program **bonap.org**. The Floristic Synthesis of North America documents more than 4,500,000 province-, state-, and county-level geographic records and includes nearly 300,000 images. It offers two distinct vegetation mapping tools, The Taxonomic Data Center (TDC) and the North American Plant Atlas (NAPA).

Center for Invasive Species and Ecosystem Health **bugwood.org**

Center for Plant Conservation **centerforplantconservation.org**. A cooperative network of thirty-six leading botanical institutions dedicated to preventing the extinction of U.S. native plants. Includes helpful resource lists and information on restoration projects.

Federal Interagency Committee for the Management of Noxious and Exotic Weeds **fs.fed.us/ficmnew/index.shtml**

Flora of North America **efloras.org**

Gymnosperm Database **conifers.org**. Provides classifications, descriptions, ecology, and uses of gymnosperms worldwide.

Invasive Plant Atlas of the United States **invasiveplantatlas.org**

Landscape Architecture Foundation Benefits Toolkit **landscapeperformance.org/benefits-toolkit**. Searchable collection of online tools and calculators for estimating landscape performance. Estimate specific landscape benefits for completed projects when actual measurements are not available or use the tools in the design phase to compare projected benefits among various options. Some of the tools also allow users to compare life-cycle costs for conventional and sustainable design features.

Missouri Botanical Garden Plant Finder **missouribotanicalgarden. org/plantfinder/plantfindersearch.aspx**. Searchable database of more than 6800 herbaceous perennials, shrubs, vines, and trees.

National Invasive Species Information Center **invasivespeciesinfo.gov**

Native Plant Information Network (Lady Bird Johnson Wildflower Center) **wildflower.org/explore**. This network strives to assemble and disseminate information that will encourage the sustainable use and conservation of native wildflowers, plants, and landscapes throughout North America. Includes horticultural and botanical information on more than 7000 species. Plant image gallery includes more than 17,000 images.

Native Plant Network **nativeplants.for.uidaho.edu**. Provides technical and practical information on growing and installing North American native plants for restoration, conservation, reforestation, landscaping, and roadside use.

Native Seed Network **nativeseednetwork.org**. Dedicated to improving the supply and management of native plant materials. Hosts an online marketplace where native seed is bought and sold, and provides an interactive ecoregion map with recommended species lists for each.

Plant Conservation Alliance **nps.gov/plants**

Plants for a Future **pfaf.org**. Research and information center focusing on edible and useful plants, alternative food crops, and conservation gardening.

U.S. Army Corps of Engineers North American Digital Flora: National Wetland Plant List **rsgisias.crrel.usace.army.mil/ NWPL/** Description of wetland regions and rating information along with a searchable list of current wetland plants for the United States.

U.S. Department of Agriculture, Natural Resources Conservation Service PLANTS database **plants.usda.gov**. The PLANTS database provides standardized information about the vascular plants, mosses, liverworts, hornworts, and lichens of the United States and Canada. Includes state checklists.

U.S. Department of Agriculture, Natural Resources Conservation Service Web Soil Survey **websoilsurvey.nrcs.usda.gov/app/ HomePage.htm**

University of Connecticut Plant Database **hort.uconn.edu** The plants listed in this resource are meant to create an awareness of the great variety of ornamental plants that will grow in USDA hardiness zone 6 or colder.

University of Michigan Dearborn Native American Ethnobotany Database **herb.umd.umich.edu**. A database of foods, drugs, dyes, and fibers of Native American peoples derived from plants.

SOURCES AND SUGGESTED READING

Ahern, J., E. Leduc, and M. L. York. 2007. *Biodiversity Planning and Design: Sustainable Practices*. Washington, D.C.: Island Press.

Anderson, M. K. 2005. *Tending the Wild: Native American Knowledge and the Management of California's Natural Resources*. Berkeley: University of California Press.

Anderson, R., J. S. Fralish, and J. M. Baskin, eds. 1999. *Savannas, Barrens, and Rock Outcrop Plant Communities of North America*. New York: Cambridge University Press.

Apfelbaum, S., and A. Haney. 2010. *Restoring Ecological Health to Your Land: A Practical Guidebook for On-the-Ground Restoration with Step-by-Step Strategies*. Washington, D.C.: Island Press.

Askins, R. 2002. *Restoring North America's Birds: Lessons from Landscape Ecology*. New Haven, CT: Yale University Press.

Bardgett, R. D. 2005. *The Biology of Soil: A Community and Ecosystem Approach*. New York: Oxford University Press.

Baskin, C. C., and J. M. Baskin. 1998. *Seeds: Ecology, Biogeography, and Evolution of Dormancy and Germination*. Waltham, MA: Academic Press.

Beck, T. 2013. *Principles of Ecological Landscape Design*. Washington, D.C.: Island Press.

Christopher, T., ed. 2011. *The New American Landscape: Leading Voices on the Future of Sustainable Gardening*. Portland, OR: Timber Press.

Clewell, A., and J. Aronson. 2008. *Ecological Restoration: Principles, Value, and Structure of an Emerging Profession*. Washington, D.C.: Island Press.

Cousens, R., and M. Mortimer. 1995. *Dynamics of Weed Populations*. New York: Cambridge University Press.

Cox, G. 2004. *Alien Species and Evolution: The Evolutionary Biology of Exotic Plants, Animals, Microbes, and Interacting Native Species*. Washington, D.C.: Island Press.

Cramer, V., and R. Hobbs. 2007. *Old Fields: Dynamics and Restoration of Abandoned Farmland*. Washington, D.C.: Island Press.

Cullina, W. 2000. *Wildflowers: A Guide to Growing and Propagating Native Flowers of North America*. Boston: Houghton Mifflin.

Cullina, W. 2002. *Native Trees, Shrubs, and Vines: A Guide to Using, Growing, and Propagating North American Woody Plants*. Boston: Houghton Mifflin.

Cullina, W. 2008. *Native Ferns, Moss, and Grasses: From Emerald Carpet to Amber Wave, Serene and Sensuous Plants for the Garden*. Boston: Houghton Mifflin.

Dale, V., and R. Haeuber, eds. 2001. *Applying Ecological Principles to Land Management*. New York: Springer.

Darke, R., and D. Tallamy. 2014. *The Living Landscape: Design for Beauty and Biodiversity in the Home Garden*. Portland, OR: Timber Press.

Dramstad, W., J. Olson, and R. T. T. Forman. 1996. *Landscape Ecology Principles in Landscape Architecture and Land-Use Planning*. Washington, D.C.: Island Press.

Dreiss, L. 2011. Differential canopy leaf flushing and site nitrogen status facilitate invasive species establishment in temperate deciduous forest understories. Master's thesis, University of Connecticut.

Dubé, R. 1997. *Natural Pattern Forms: A Practical Sourcebook for Landscape Design*. New York: Van Nostrand Reinhold.

Dunnett, N., and J. Hitchmough, eds. 2004. *The Dynamic Landscape: Ecology, Design, and Management of Naturalistic Urban Planting*. Philadelphia: Taylor and Francis.

Egan, D., and E. Howell, eds. 2005. *The Historical Ecology Handbook: A Restorationist's Guide to Reference Ecosystems*. Washington, D.C.: Island Press.

Egler, F. 1977. *The Nature of Vegetation*. Norfolk, CT: Aton Forest.

Grese, R., 2011. *The Native Landscape Reader*. Amherst, MA: Library of American Landscape History.

Gurevitch, J., S. Scheiner, and G. Fox. 2006. *The Ecology of Plants*. 2nd ed. Sunderland, MA: Sinauer.

Harker, D., et al. 1999. *Landscape Restoration Handbook*. 2nd ed. Boca Raton, FL: Lewis.

Hartman, J. M., J. F. Thorne, and C. E. Bristow. 1993. Variations in old field succession. *Council of Educators in Landscape Architecture Selected Papers* 4:55–62.

Hebberger, J. 2000. The effects of soil amendments on mycorrhizae and growth of *Quercus rubra* (northern red oak) seedlings. Ph.D. dissertation, Rutgers University.

Hightshoe, G. 1987. *Native Trees, Shrubs, and Vines of Urban and Rural America: A Planting Design Manual for Environmental Design*. Hoboken, NJ: Wiley.

Holm, H. 2013. Milkweed pollination: A sticky situation. Restoring the Landscape with Native Plants web site. restoringthelandscape.com/2013/02/milkweed-pollination-sticky-situation.html

Holm, H. 2014. *Pollinators of Native Plants: Attract, Observe and Identify Pollinators and Beneficial Insects with Native Plants*. Minnetonka, MN: Pollination Press.

Howard, V., and P. Robins. 2002. *Capay Valley Conservation and Restoration Manual: A Handbook for Landowners*. yolorcd.org/documents/cv_conservation_restoration_manual.pdf.

Kaplan, R., S. Kaplan, and R. Ryan. 1998. *With People in Mind: Design and Management of Everyday Nature*. Washington, D.C.: Island Press.

Kaufman, S. R., and W. Kaufman. 2007. *Invasive Plants: Guide to Identification and the Impacts and Control of Common North American Species*. Mechanicsburg, PA: Stackpole.

Kenfield, W. 1991, ed. *The Wild Gardener in the Wild Landscape: The Art of Naturalistic Landscaping*. New London, CT: Connecticut College Arboretum.

Kowarick, I., and S. Korner, eds. 2005. *Wild Urban Woodlands: New Perspectives in Urban Forestry*. New York: Springer.

Kress, S. W. 2006. *The Audubon Society Guide to Attracting Birds*. Ithaca, NY: Cornell University.

Kurtz, C. 2001. *Practical Guide to Prairie Reconstruction*. Iowa City: University of Iowa Press.

Luken, J. O. 1990. *Directing Ecological Succession*. New York: Springer.

Nicholson, R. 2011. Little big plant, box huckleberry (*Gaylussacia brachycera*). *Arnoldia* 68:11–18.

Packard, S., and C. F. Mutel, eds. 2005. *The Tallgrass Restoration Handbook: For Prairies, Savannas, and Woodlands*. Washington, D.C.: Island Press.

Phillips H. R. 1985. *Growing and Propagating Wild Flowers*. Chapel Hill: University of North Carolina Press.

Pickett, S., and P. S. White. 1985. *The Ecology of Natural Disturbance and Patch Dynamics*. Waltham, MA: Academic Press.

Roberts, E., and E. Rehmann, eds. 1996. *American Plants for American Gardens*. Athens: University of Georgia Press.

Russell, E. W. B. 1998. *People and Land through Time: Linking Ecology and History*. New Haven, CT: Yale University Press.

Sauer, L. 1998. *The Once and Future Forest: A Guide to Forest Restoration Strategies*. Washington, D.C.: Island Press.

Soderstrom, N. 2008. *Deer-Resistant Landscaping: Proven Advice and Strategies for Outwitting Deer and 20 Other Pesky Mammals*. New York: Rodale.

Stein, S. B. 1993. *Noah's Garden: Restoring the Ecology of Our Own Back Yards*. Boston: Houghton Mifflin.

Tallamy, D. 2009. *Bringing Nature Home: How You Can Sustain Wildlife with Native Plants*. Portland, OR: Timber Press.

Thoren, R. 2014. *Landscapes of Change: Innovative Designs and Reinvented Sites*. Portland, OR: Timber Press.

Tiner, R. 2005. *In Search of Swampland: A Wetland Sourcebook and Field Guide*. New Brunswick, NJ: Rutgers University Press.

U.S. Department of Agriculture Natural Resources Conservation Service and Missouri Department of Conservation. 2006. *Central Region Seedling ID Guide for Native Prairie Plants*. USDA-NRCS and Missouri Department of Conservation.

U.S. Environmental Protection Agency. 2013. Level III ecoregions of the continental United States. epa.gov/wed/pages/ecoregions/na_eco.htm.

Watson, W. T. 2005. Influence of tree size on transplant establishment and growth. *HortTechnology* 15:118–122.

Wessels, T. 2010. *Forest Forensics: A Field Guide to Reading the Forested Landscape*. Woodstock, VT: Countryman.

Williams, D. 2010. *The Tallgrass Prairie Center Guide to Seed and Seedling Identification*. Iowa City: University of Iowa Press.

Woelfle-Erskine, C., and A. Uncapher. 2012. *Creating Rain Gardens: Capturing the Rain for Your Own Water-Efficient Garden*. Portland, OR: Timber Press.

Young, J. A., and C. G. Young. 1986. *Collecting, Processing, and Germinating Seeds of Wildland Plants*. Portland, OR: Timber Press.

Young, J. A., and C. G. Young. 2009. *Seeds of Woody Plants in North America*. Rev. ed. Portland, OR: Timber Press.

ACKNOWLEDGMENTS

This book could not have been completed without the outstanding contributions of my design colleague Jenna Webster. Her attention to detail, artistic contributions, and intellectual input made this book infinitely better.

It is entirely possible that my career would not be where it is today had it not been for Luther Van Ummersen, a true friend and one of the most unselfish people I have ever met.

I thank everyone in our office and extended family, including Kate Butler, Rebecca Kagle, Wayne Lee, and Pennington Marchael, whose competence, dedication, and companionship make our firm better and my work fun. I also thank Jerry Albert, Jim Barrett, Brian O'Neill, Larry Rossi, Ludin Sandoval, Nery Sandoval, and Jeff Schumacher, who have all generously contributed their expertise and friendship.

I thank Ian Caton, design associate and friend, whose knowledge of plants will never cease to amaze me; Glenn Dreyer, friend and colleague, who brought me more deeply into the world of ecological science; and Eugene Varady, my first employer in the garden world, whose iconoclastic vision influenced me tremendously in my early years and beyond.

Finally, I thank all of my clients, who have allowed me to follow my muse where they live and work and have generously offered me their hospitality and friendship through the years.

—LARRY WEANER

Thanks (as always) to Suzanne O'Connell for her support and to Ruth Clausen for inspiration and wise advice.

—TOM CHRISTOPHER

PHOTOGRAPHY AND ILLUSTRATION CREDITS

PHOTOGRAPHY

Karen Bussolini, pages 10–11, 52, 53 top, 55, 78, 81 bottom right, 116 third of four, 117, 148, 158–159, 187, 255, 258

A. E. Bye, courtesy Penn State University Archives, The Pennsylvania State University Libraries, page 30

Rob Cardillo, cover and pages 6, 9, 12, 19, 27, 29, 48, 56, 69, 80, 81 bottom left, 84, 91, 92, 125, 126, 133, 146 right–147, 170, 173, 185, 200, 203 top, 204, 205 bottom, 256, 287, 297, 298, 299, 300, 302, 313 bottom

Ian Caton, pages 91, 120, 121, 146 left, 152, 155, 160, 161, 212 top, 233 bottom, 243 first column middle, second column top and bottom, third column top

William Cullina, page 60

J. Wayne Lee Jr., page 154

Patricia Drackett, page 220 top

Bruce Leander, Lady Bird Johnson Wildflower Center, page 219

Hawthorne Valley Farmscape Ecology Program, pages 212 bottom, 236 bottom

David Korbonits, page 220 bottom

Clark Montgomery, page 47

Linda Weaner, page 37

Mark Weaner, pages 99, 164 bottom, 165, 166, 167, 168, 169 top

Jenna Webster, pages 264 top left and bottom, 265, 291

Thena Webster, page 59

All other photographs are by Larry Weaner.

ILLUSTRATIONS

Anna Eshelman based on drawings by Larry Weaner Landscape Associates, pages 189, 193, 271, 288

Larry Weaner and Ian Caton, page 141

Larry Weaner Landscape Associates, pages 97, 98, 101, 104, 109

INDEX